Structurally Constrained Controllers

T0143089

Somayeh Sojoudi • Javad Lavaei
Amir G. Aghdam

Structurally Constrained Controllers

Analysis and Synthesis

Springer

Somayeh Sojoudi
California Institute of Technology
Control & Dynamical Systems Dept.
Mail Code 107-81
91125 Pasadena California
USA
sojoudi@cds.caltech.edu

Javad Lavaei
California Institute of Technology
Control & Dynamical Systems Dept.
Mail Code 107-81
91125 Pasadena California
USA
lavaei@cds.caltech.edu

Amir G. Aghdam
Concordia University
Dept. Electrical & Computer Engineering
1455 de Maisonneuve Blvd. West
EV005.139
Montreal Québec H3G 1M8
Canada
aghdam@ece.concordia.ca

ISBN 978-1-4899-8542-2 ISBN 978-1-4419-1549-8 (eBook)
DOI 10.1007/978-1-4419-1549-8
Springer New York Dordrecht Heidelberg London

Printed on acid-free paper

Springer is part of Springer Science+Business Media (www.springer.com)

...dedicated to

my mother —*from* Somayeh Sojoudi

my parents —*from* Javad Lavaei

my wife —*from* Amir G. Aghdam

Preface

Decentralized control of large-scale systems has been an active area of research in the past 40 years. The need for computationally efficient strategies to control high-order systems with a large number of inputs and outputs was the initial motivation for extensive research in this area. However, due to significant advances in computer technology, this may not be the main concern in practice. On the other hand, in more recent practical applications of large-scale control systems (such as cooperative control of autonomous vehicles), reliability is one of the most important design specifications; and of course reliability is a clear advantage of decentralized control structure over a centralized one because unlike a centralized structure, there is no single point of failure in a decentralized control configuration. Many universities in the world have a course entitled "large-scale control systems" in their graduate curriculum, where some classical results in decentralized control are covered. However, the need for a reference which addresses more recent problems in this area is clear. On one hand, the introduction of the new wave of applications of decentralized control in the past decade has provided a new dimension to this area of research. On the other hand, the rapid growth in computational power and resources has necessitated a major change in the composition of a systematic course in this area. The classical results in theory of decentralized large-scale systems build the foundation of the course contents, and the practical applications are important requirements of a meaningful engineering education, which makes the course more attractive from the student's viewpoint. And these two components concurrently define effective teaching for a graduate course.

To the best of the authors' knowledge, there is no book in the market which presents recent problems and methodologies in decentralized control of systems, without neglecting the important concepts introduced in earlier years, when the classical results in this area were developed. Therefore, professors often use a collection of relevant papers to address more recent research results and applications throughout the course. This book is an attempt to fill the gap between the classical results in decentralized control of large-scale interconnected systems developed in 1970's and 80's, and some important problems which were introduced by more recent applications of decentralized control, with a special focus on multi-agent systems application. These problems include robust stability and stabilizability, simplest control structure required for stability, performance analysis, and information exchange between the control components. The authors hope that this can be a good reference book for any course in the area of large-scale control systems, and that the students will find it useful in identifying some open problems in this field of research.

Contents

1 **Introduction** ... 1

2 **Characterization of Decentralized and Quotient Fixed Modes Via Graph Theory** ... 9
2.1 Introduction ... 9
2.2 Preliminaries ... 10
2.3 Characterization of decentralized fixed modes ... 13
2.4 Characterization of quotient fixed modes ... 16
2.5 Illustrative examples ... 17
2.6 Summary ... 19
References ... 21

3 **Time Complexity of Decentralized Fixed Mode Verification** ... 23
3.1 Introduction ... 23
3.2 Preliminaries and problem formulation ... 25
3.2.1 Matrix rank checking ... 26
3.2.2 Randomized algorithm ... 26
3.2.3 Derandomization ... 26
3.2.4 Graph-theoretic approach ... 27
3.2.5 Problem statement ... 28
3.3 Main results ... 28
3.4 Numerical example ... 31
3.5 Summary ... 32
3.6 Appendix ... 33
References ... 34

4 **Decentralized Overlapping Control: Stabilizability and Pole-Placement** ... 37
4.1 Introduction ... 37
4.2 Problem formulation ... 38
4.3 Computing the transformation matrices ... 40

4.4 Linear time-invariant control law ... 45
4.4.1 Pole placement ... 45
4.4.2 Optimal LTI controller ... 48
4.5 Non-LTI control law ... 49
4.5.1 Generalized sampled-data hold function ... 50
4.5.2 Sampled-data controller ... 51
4.5.3 Finite-dimensional linear time-varying controller ... 52
4.5.4 General controller ... 53
4.6 Comparison with existing methods ... 55
4.6.1 Comparison with the work presented in [8] ... 55
4.6.2 Comparison with the work presented in [25] ... 56
4.7 Numerical examples ... 56
4.8 Summary ... 59
References ... 59

5 Elimination of Decentralized Fixed Modes via Optimal Information Exchange ... 61
5.1 Introduction ... 61
5.2 Preliminaries ... 63
5.3 Main Results ... 64
5.3.1 Displacing a single unrepeated DFM ... 64
5.3.2 Displacing multiple unrepeated DFMs ... 68
5.4 Numerical example ... 69
5.5 Summary ... 71
References ... 71

6 Characterization of Stabilizing Structurally Constrained Controllers ... 73
6.1 Abstract ... 73
6.2 Introduction ... 73
6.3 Problem formulation ... 75
6.4 Main results ... 77
6.4.1 An efficient algorithm to obtain the optimal control interaction set(s) ... 81
6.5 Numerical example ... 82
6.6 Summary ... 85
References ... 86

7 LQ Decentralized Controllers with Disturbance Rejection Property for Hierarchical Systems ... 87
7.1 Introduction ... 87
7.2 Problem formulation ... 89
7.3 Preliminaries ... 91
7.4 A reference centralized servomechanism controller ... 92
7.5 Optimal decentralized servomechanism controller ... 96

7.6 Practical considerations in control design 99
7.7 Numerical examples 101
7.8 Summary 106
References 106

8 Decentralized Implementation of Centralized Controllers for Interconnected Systems 109
8.1 Introduction 109
8.2 Problem formulation 111
8.3 Main results 112
8.3.1 Structural initial value observer 118
8.3.2 Decentralization of centralized controllers 121
8.3.3 Optimal decentralized controller 122
8.4 Numerical example 125
8.5 Summary 127
8.6 Appendix 1 127
8.6.1 Static centralized controllers 128
8.6.2 Extension to dynamic centralized controllers 134
8.7 Appendix 2 135
References 136

9 Robust Control of LTI Systems by Means of Structurally Constrained Controllers 139
9.1 Introduction 139
9.2 Robustness property of the modes of a LTI system 141
9.3 Numerical example 147
9.4 Summary 149
References 149

10 Robust Control of Large-Scale Systems: Efficient Selection of Inputs and Outputs 151
10.1 Introduction 151
10.2 Preliminaries and problem formulation 153
10.2.1 Background on Sum-of-Squares 154
10.3 Robust controllability degree 155
10.3.1 Special case: A polytopic region 160
10.3.2 Comparison with existing results 164
10.4 Numerical example 164
10.5 Summary 167
References 169

11 Interconnection-Based Performance Analysis for a Class of Decentralized Controllers 171
11.1 Introduction 171
11.2 Preliminaries and problem formulation 173
11.3 Main results 176

11.4 Numerical example 181
11.5 Summary 185
11.6 Appendix 185
References 187

Index 189

Chapter 1
Introduction

In the control literature, a system consisting of several interconnected subsystems (which may be geographically distributed) is referred to as a large-scale system. In this type of system, it is often desired to perform the control function in a decentralized fashion. In fact, for such systems it is not realistic to assume that all output measurements can be transmitted to every local control station. Problems of this kind appear for example in electric power systems, communication networks, large space structures, robotic systems, economic systems and traffic networks, to name only a few. Typical large-scale control systems have several local control stations, which observe only local outputs and control only local inputs. Every local controller is involved, however, in the operation of the overall system.

Consider the temperature control system for a building with multiple floors and several rooms on each floor. This is an example of an interconnected system where the temperature of each room affects the temperature of the other rooms in the building as well. One approach to control this system is to collect the measured temperature of all rooms, and process the information in a centralized fashion to compute the temperature control signal for each room. As a decentralized alternative, one can use the information about the interconnection dynamics of the entire system to compute the temperature control signal for each room by using only the measured temperature of that room (along with the temperature of the neighboring rooms, at most). Each of these two control structures has advantages of disadvantages. The main advantage of a centralized control structure is that its performance is in general better than that of a decentralized structure. However, this improved performance is at a high price. First of all, the computational requirement in a centralized control structure is much higher than that in a decentralized structure, as each control signal is computed based on a large amount of information, in general, while most of the non-local information may have a negligible impact on the corresponding control signal. Moreover, a centralize controller has a single point of failure, while the failure of a local controller in a decentralized structure does not necessarily lead to the failure of the entire control system.

Another motivating example is the formation flying control of a number of spacecraft in order to accomplish a mission cooperatively. The structure of information exchange between the spacecraft determines the architecture of the formation, which can be classified in five main categories: leader-follower, behavioral, virtual, cyclic, and multi-input multi-output. This classification is, in fact, based on the topology of communication between the spacecraft controllers. In practice, it is preferred to have the minimum number of communication links, as communication is very expensive in deep space. Insufficient information exchange, on the other hand, may causes several problems such as deterioration of the overall control performance, inability to avoid collision, inability to detect obstacles, and inefficient formation reconfiguration. The design of a structured cooperative control requiring the minimum number of communication links in order to achieve certain performance objectives is one of the recent topics studied in the area of decentralized control.

The objective of this book is to use control theory to design and analyze decentralized feedback control laws for large-scale systems. Some important results in this area will be reviewed, and more efficient alternative approaches will also be provided in some cases. As an important example of the recent applications of decentralized control, cooperative control of multi-agent systems is investigated in a number of chapters, and some important issues such as stability, robustness, and control performance are addressed. A description of the material covered in each chapter of this book is spelled out in the sequel.

Chapter 2 deals with the decentralized control of systems with distinct modes. A simple graph-theoretic approach is first proposed to identify those modes of the system which cannot be moved by means of a linear time-invariant decentralized controller. To this end, the system is transformed into its Jordan state-space representation. Then, a matrix is computed, which has the same order as the transfer function matrix of the system. A bipartite graph is constructed from the computed matrix. Using this approach, the problem of characterizing the decentralized fixed modes (DFM) of the system reduces to verifying if the above graph has a complete bipartite subgraph with a certain property. Analogously, a graph-theoretic method is presented to compute the modes of the system which are fixed with respect to any general (nonlinear and time-varying) decentralized controller. The proposed approaches are much simpler than the existing ones, which typically require calculating the ranks of several matrices.

Given an interconnected system, Chapter 3 is concerned with the time complexity of verifying whether an unrepeated mode of the system is a decentralized fixed mode. It is shown that checking the decentralized fixedness of any distinct mode is tantamount to testing the strong connectivity of a digraph obtained from the system. It is subsequently proved that the corresponding decision problem is of the same time-complexity as matrix multiplication. This chapter concludes that the identification of distinct decentralized fixed modes (by means of a deterministic algorithm, rather than a randomized one) is computationally very easy, although the existing algorithms for solving this problem would wrongly imply that it is cumbersome. This chapter provides the complexity analysis as well as an efficient algorithm for tackling the underlying problem.

Chapter 4 deals with the control of large-scale interconnected systems with a constrained control structure. It is shown that certain modes of the system can be freely placed anywhere in the complex plane, by using a linear time-invariant (LTI) structurally constrained controller. These modes have been identified by introducing the notion of a decentralized overlapping fixed mode (DOFM). This implies that the system is stabilizable by a LTI structurally constrained controller if and only if it does not have any unstable DOFM. Furthermore, a design procedure is proposed for obtaining a controller to achieve the desired pole placement for the systems with no DOFM. In addition, the problem of designing a structurally constrained optimal LTI controller with respect to a quadratic performance index is studied. Designing various types of structurally constrained controllers (such as periodic feedback) is then investigated. The notion of a quotient overlapping fixed mode (QOFM) is also introduced, and it is shown that a system is stabilizable by mean of a general controller, i.e. nonlinear and time-varying, if and only if it does not have any unstable QOFM. In the case of no unstable QOFM, it is proved that there exists a finite-dimensional linear time-varying structurally constrained controller to stabilize the system.

Chapter 5 is concerned with the stabilizability of interconnected systems via LTI decentralized controllers. Given a system with some distinct DFMs, it is desired to find a desirable control structure (in terms of information flow) for it. Since a decentralized controller consists of a number of non-interacting local controllers, the objective here is to establish certain interactions between the local controllers in order to eliminate the undesirable DFMs. The resultant control configuration in the presence of new interactions will be overlapping (as a more general form of decentralized control). In other words, this chapter characterizes all decentralized overlapping control structures with respect to which none of the undesirable modes are fixed. This objective is achieved by translating the knowledge of the system into some bipartite graphs. Then, the notions of minimal sets and maximal subgraphs are introduced, which lead to a simple combinatorial algorithm for solving the underlying problem. The efficacy of the results obtained is demonstrated in an illustrative example.

The focus of Chapter 6 is directed towards the problem of characterizing the information flow structures of all classes of LTI structurally constrained controllers with respect to which a given interconnected system has no fixed modes. Any class of structurally constrained controllers can be described by a set of communication links, which delineates how the corresponding local controllers interact with each other. To achieve the objective, a cost is first attributed for establishing any communication link in the control structure. These costs are part of design specification and represent the expenditure of data transmission between different subsystems. A simple graph-theoretic method is then proposed to characterize all the relevant classes of controllers systematically. As a by-product of this approach, all classes of LTI stabilizing structurally constrained controllers with the minimum implementation cost are attained using the algorithm provided in this chapter. The primary advantages of this approach are its simplicity and computational efficiency. The efficacy and importance of the results of this chapter are thoroughly illustrated by a numerical example. This chapter integrates the ideas proposed in a recently published work and some original techniques to develop its main results.

Chapter 7 is concerned with decentralized output regulation of hierarchical systems subject to input and output disturbances. It is assumed that the disturbance can be represented as the output of an autonomous LTI system with unknown initial state. The primary objective is to design a decentralized controller with the property that not only does it reject the degrading effect of the disturbance on the output (for a satisfactory steady-state performance), it also results in a small linear quadratic (LQ) cost function (implying a good transient behavior). To this end, the underlying problem is treated in two phases. In the first step, a number of modified systems are defined in terms of the original system. The problem of designing a LQ centralized controller which stabilizes all the modified systems and rejects the disturbance in the original system is considered, and it is shown that this centralized controller can be efficiently found by solving a linear matrix inequality (LMI) problem. In the second step, a method recently presented in the literature is used to decentralize the designed centralized controller. It is proved that the obtained controller satisfies the prescribed design specifications including disturbance rejection. In order to put the results into a more pragmatic framework, the system is then assumed to be subject to input delays, and a robustness analysis is carried out accordingly. Simulation results elucidate the efficacy of the proposed control law.

Given a centralized controller associated with a LTI interconnected system, Chapter 8 aims to design a parameterized decentralized controller such that the state and input of the system under the obtained decentralized controller can become arbitrarily close to those of the system under the given centralized controller, by tuning the controller's parameters. To this end, a two-level decentralized controller is designed, where the upper level captures the dynamics of the centralized closed-loop system, and the lower level is an observer-based sub-controller designed based on the new notion of structural initial value observability. This method can be used to decentralize every generic centralized controller, provided the interconnected system satisfies very mild conditions. The efficacy of results is elucidated by some numerical examples.

Chapter 9 investigates the stabilizability property of uncertain systems via structurally constrained controllers. First, a LTI uncertain system is considered whose state-space matrices depend polynomially on the uncertainty vector, defined over a region. It is shown that if the system is stabilizable by a structurally constrained controller in one point belonging to the region, then it is stabilizable by a controller with the same structural constraint in all points pertaining to the region, with possible exception of some points located on an algebraic variety. Thus, if a system is stabilizable via a constrained controller at the nominal point, then it is almost always stabilizable at any operating point around the nominal model. It is also shown how this algebraic variety (or the dominant subvariety of it) can be computed efficiently. This chapter provides more general results compared to the existing ones in the literature for the robust stability of the LTI systems.

A practical approach to control a large-scale system efficiently is to design a structurally constrained robust controller which employs a minimum number of inputs and outputs (or the least-expensive combination of them). Such an approach consists of two phases: first, it is required to contrive the desired structure (topology) of the controller (referred to as information flow structure), and then the controller parameters are to be found. Once the first phase is completed, the second step can be accomplished using the existing methods in the literature. Chapter 10 is concerned with finding a proper information flow structure, i.e., the first step in designing a structurally constrained controller for an interconnected system (as noted above). To this end, it is assumed that the system is polynomially uncertain over a known semi-algebraic set. Given a control topology, it is desired to measure the robust controllability/observability of the system with respect to this control structure. A sum-of-squares (SOS) optimization is provided to minimize the smallest singular value of the controllability or observability Gramian over the uncertainty region. This provides a quantitative measure for the robust controllability or observability degree of the system. In the special case when the uncertainty region is polytopic, the corresponding SOS formulation can be simplified significantly. One can apply the proposed method to any large-scale interconnected system in order to identify those inputs and outputs that are more effective in controlling the system, in a robust manner. This enables the designer to simplify the control structure by ignoring those inputs and outputs whose contribution to the overall control operation is relatively weak. A numerical example is presented to demonstrate the efficacy of the results.

Chapter 11 deals with decentralized controller design for large-scale interconnected systems of pseudo-hierarchical structure. Given such a system, one can use existing techniques to design a decentralized controller for a "reference" hierarchical model, obtained by eliminating certain weak interconnections of the original system. Although this indirect controller design is appealing as far as the computational complexity is concerned, it does not necessarily result in satisfactory performance for the original pseudo-hierarchical system. A LQ cost function is defined in order to evaluate the performance discrepancy between the pseudo-hierarchical system and its reference hierarchical model under the designed decentralized controller. A discrete Lyapunov equation is then solved to compute this performance index. However, due to the large-scale nature of the system, this equation cannot be handled efficiently, in general. Thus, attaining an upper bound on this cost function can be more desirable than finding its exact value, in practice. For this purpose, a novel technique is proposed which only requires solving a simple LMI optimization problem with three variables. The problem is then reduced to a scalar optimization problem, for which an explicit solution is provided. It is also shown that when the original model is exactly hierarchical, the upper bounds obtained from the LMI and scalar optimization problems will both be equal to zero.

Some of the materials of this book have been reprinted from the following journal and conference papers (with permission from the corresponding copyright owners):

- J. Lavaei, "Decentralized implementation of centralized controllers for interconnected systems," accepted (subject to revisions) in *IEEE Transactions on Automatic Control*, 2010.

- J. Lavei and S. Sojoudi, "Time complexity of decentralized fixed mode verification, *IEEE Transactions on Automatic Control*, vol. 55, no. 4, pp. 971-976, 2010.
- S. Sojoudi and A. G. Aghdam, "Interconnection-based performance analysis for a class of decentralized controllers," *Automatica*, vol. 46, Issue 5, pp. 796-803, 2010.
- J. Lavaei and A. G. Aghdam, "Performance improvement of robust controllers for polynomially uncertain systems, *Automatica*, vol. 46, vol. 1, pp. 110-115, 2010.
- S. Sojoudi, J. Lavaei and A. G. Aghdam, "Robust controllability and observability degrees of polynomially uncertain systems, *Automatica*, vol. 45, no. 11, pp. 2640-2645, 2009.
- J. Lavaei and A. G. Aghdam, "Overlapping control design for multi-channel systems, *Automatica*, vol. 45, no. 5, pp. 1326-1331, 2009.
- S. Sojoudi and A. G. Aghdam, "Overlapping control systems with optimal information exchange," *Automatica*, vol. 45, no. 5, pp. 1176-1181, 2009.
- J. Lavaei and A. G. Aghdam, "Control of continuous-time LTI systems by means of structurally constrained controllers," *Automatica*, vol. 44, no. 1, 2008.
- J. Lavaei, A. Momeni and A. G. Aghdam, "LQ suboptimal decentralized controllers with disturbance rejection property for hierarchical systems", *International Journal of Control*, vol. 81, no. 11, pp. 1720-1732, 2008.
- J. Lavaei and A. G. Aghdam, "A graph theoretic method to find decentralized fixed modes of LTI systems," *Automatica*, vol. 43, no. 12, pp. 2129-2133, 2007.
- S. Sojoudi, J. Lavaei and A. G. Aghdam, "Robust stabilizability verification of polynomially uncertain LTI systems," *IEEE Transactions on Automatic Control*, vol. 52, no. 9, pp. 1721-1726, 2007.
- J. Lavaei, "A new decentralization technique for interconnected systems," in *Proceedings of 48th IEEE Conference on Decision and Control*, Shanghai, China, 2009.
- S. Sojoudi, J. Lavaei and A. G. Aghdam, "Controllability and observability of uncertain systems: a robust measure," in *Proceedings of 48th IEEE Conference on Decision and Control*, Shanghai, China, 2009.
- J. Lavaei and S. Sojoudi, "Complexity of checking the existence of a stabilizing decentralized controller," in *Proceedings of 2009 American Control Conference*, St. Louis, MO, 2009.
- J. Lavaei and A. G. Aghdam, "LQ robust controllers for polynomially uncertain systems," in *Proceedings of 47th IEEE Conference on Decision and Control*, Cancun, Mexico, 2008.
- S. Sojoudi and Amir G. Aghdam, "Performance analysis for a class of decentralized control systems," in *Proceedings of 2008 American Control Conference*, Seattle, WA, pp. 710-716, 2008.
- J. Lavaei, Ahmadreza Momeni and A. G. Aghdam, "A near-optimal decentralized servomechanism controller for hierarchical interconnected systems," in *Proceedings of 46th IEEE Conference on Decision and Control*, New Orleans, LA, 2007.

- J. Lavaei and A. G. Aghdam, "Characterization of decentralized and quotient fixed modes via graph theory," in *Proceedings of 2007 American Control Conference*, New York, NY, 2007.
- J. Lavaei and A. G. Aghdam, "On structurally constrained control design with a prespecified form," in *Proceedings of 2007 American Control Conference*, New York, NY, 2007.
- J. Lavaei, S. Sojoudi and A. G. Aghdam, "Stability criterion for general proper systems with constrained control structure," in *Proceedings of 2007 American Control Conference*, New York, NY, 2007.
- S. Sojoudi, J. Lavaei and A. G. Aghdam, "Optimal information flow structure for control of interconnected systems," in *Proceedings of 2007 American Control Conference*, New York, NY, 2007.
- S. Sojoudi and A. G. Aghdam, "Characterizing all classes of LTI stabilizing structurally constrained controllers by means of combinatorics," in *Proceedings of 46th IEEE Conference on Decision and Control*, New Orleans, LA, 2007.

Chapter 2
Characterization of Decentralized and Quotient Fixed Modes Via Graph Theory

2.1 Introduction

Many real-world systems can be envisaged as the interconnected systems consisting of a number of subsystems. Normally, the desirable control structure for this class of systems is decentralized, which comprises a set of local controllers for the subsystems [1, 2, 3, 4, 5, 6]. Decentralized control theory has found applications in large space structure, communication networks, power systems, etc. [7, 8, 9, 10]. More recently, simultaneous stabilization of a set of decentralized systems and decentralized periodic control design are investigated in [11, 12].

The notion of a decentralized fixed mode (DFM) was introduced in [1], where it was shown that any mode of a system which is not a DFM can be placed freely in the complex plane by means of an appropriate linear time-invariant (LTI) controller. An algebraic characterization of DFMs was presented in [13]. A method was then proposed in [14] to characterize the DFMs of a system in terms of its transfer function. It was shown in [15] that the DFMs of any system can be attained by computing the transmission zeros of a set of systems derived from the original system. In [2], an algorithm was presented to identify the DFMs of the system by checking the rank of a set of matrices. It is worth noting that the number of the systems whose transmission zeros need to be checked in [15] and the number of matrices whose ranks are to be computed in [2] depend exponentially on the number of the subsystems of the original system. This means that while these methods are theoretically developed for any multi-input multi-output (MIMO) system, they are computationally ill-conditioned. The method introduced in [16] addresses this shortcoming by partitioning the system into a number of modified subsystems, obtained based on the strong connectivity of the system's graph. Then, instead of finding the DFMs of the original system, one can compute the DFMs of the modified subsystems to reduce the corresponding computational complexity. However, the computational burden can still be high when the system consists of several strongly connected subsystems. In general, the method given in [16] is more effective for medium-sized systems, while the one in [2] is only appropriate for small-sized systems. It is to be noted that the method introduced in [2] is widely used in the literature for the characterization of the DFMs.

On the other hand, the notion of a quotient fixed mode (QFM) is introduced in [17] to identify those modes of the system which are fixed with respect to general (nonlinear and time-varying) decentralized controllers. The properties of QFM is further investigated in [3], where it is asserted that the non-quotient DFMs of a broad class of systems can by eliminated by means of sampling.

This chapter aims to present simple approaches to find the DFMs and the QFMs of a system with distinct modes. To this end, a matrix is obtained first, which resembles the transfer function matrix of the system at one point. Then, a bipartite graph is constructed in terms of this matrix. It is shown that having a complete bipartite subgraph with a certain property is equivalent to having a DFM. A similar method is pursued to obtain the QFMs of the system. The combinatorial approaches proposed in this chapter are substantially simpler than the conventional methods for finding the DFMs and QFMs. The efficacy of the proposed methods is demonstrated in two numerical examples.

2.2 Preliminaries

Consider a LTI interconnected system $\mathscr{S}$ consisting of ν subsystems $S_1, S_2, ..., S_\nu$. Denote the modes of the system with $\sigma_1, \sigma_2, ..., \sigma_n$, and assume that all of them are distinct. The system $\mathscr{S}$ can be realized in the decoupled form as:

$$\dot{x}(t) = Ax(t) + \sum_{j=1}^{\nu} B_j u_j(t)$$

$$y_i(t) = C_i x(t) + \sum_{j=1}^{\nu} D_{ij} u_j(t), \quad i \in \bar{\nu} := \{1, 2, ..., \nu\}$$

where $x(t) \in \Re^n$ is the state, and $u_i(t) \in \Re^{m_i}$ and $y_i(t) \in \Re^{r_i}$, $i \in \bar{\nu}$, are the input and the output of the ith subsystem, respectively, and:

$$A = \text{diag}\left(\left[\,\sigma_1\ \sigma_2\ \cdots\ \sigma_n\,\right]\right) \tag{2.1}$$

Throughout this chapter, the term "decentralized controller" is referred to the union of all local controllers. In order to specify the local subsystems associated with the local controllers, the subsystems are enclosed within parentheses throughout this chapter, if necessary. For instance, a decentralized controller for the system $\mathscr{S}(S_1, S_2, S_3)$ is the union of the local controllers $u_i(t) = g_i(y_i(t), t)$, $i \in \{1, 2, 3\}$, corresponding to the subsystems S_1, S_2, S_3. Some of the important notions for different types of fixed modes will be given next.

Definition 1 ([1]) $\lambda \in sp(A)$ *is said to be a decentralized fixed mode (DFM) of the system* $\mathscr{S}$*, if it remains a mode of the closed-loop system under any arbitrary* *decentralized static feedback. In other words, $\lambda \in sp(A)$ is a DFM of the system $\mathscr{S}$ if and only if:*

$$\lambda \in sp\left(A+\sum_{i=1}^{v} B_iK_iC_i\right), \quad \forall K_i \in \Re^{m_i \times r_i},\ i \in \bar{v} \tag{2.2}$$

It can be shown that a DFM is fixed with respect to any arbitrary dynamic LTI decentralized controller. However, it is interesting to note that a proper non-LTI controller can eliminate certain types of DFMs [3]. In other words, a DFM is not necessarily fixed with respect to a time-varying or nonlinear control structure.

Definition 2 *Define the structural graph of the system $\mathscr{S}$ as a digraph with v vertices which has a directed edge from the ith vertex to the jth vertex if $C_j(sI-A)^{-1}B_i \neq 0$, for any $i, j \in \bar{v}$. The structural graph of the system $\mathscr{S}$ is denoted by $\mathscr{G}$.*

Partition $\mathscr{G}$ into the minimum number of strongly connected subgraphs denoted by $G_1, G_2, \ldots., G_l$ (note that a digraph is called strongly connected iff there exists a directed path from any vertex to any other vertices of the graph [17, 3]). Define the subsystem $\tilde{S}_i$, $i = 1,2,\ldots,l$, as the union of all subsystems of $\mathscr{S}$ corresponding to the vertices in the subgraph G_i (note that vertex j in the graph $\mathscr{G}$ represents the subsystem S_j, for any $j \in \bar{v}$).

Definition 3 ([17]) *Assume that the system $\mathscr{S}$ is strictly proper, i.e. $D = 0$. The mode λ is said to be a quotient fixed mode (QFM) of the system $\mathscr{S}(S_1, S_2, \ldots, S_v)$, if it is a DFM of the system $\mathscr{S}(\tilde{S}_1, \tilde{S}_2, \ldots, \tilde{S}_l)$.*

It can be shown in [17] that a QFM is fixed with respect to any arbitrary (nonlinear or time-varying) decentralized controller. In order to clarify the notion of QFMs, consider the system $\mathscr{S}$ with the parameters given below:

$$\begin{aligned} A &= \text{diag}([1\ ,\ -2\ ,\ -3]),\ B_1 = [0\ \ 0\ -1]^T, \\ B_2 &= [1\ \ 1\ \ 2]^T,\ B_3 = [2\ \ 1\ \ 5]^T, \\ C_1 &= [5\ \ 3\ \ 2],\ C_2 = [0\ -1\ \ 0],\ \ C_3 = [0\ -2\ \ 0] \end{aligned} \tag{2.3}$$

The transfer function matrix of this system will be equal to:

$$C(sI-A)^{-1}B = \begin{bmatrix} -2s-6 & 12s+13 & 23s+26 \\ 0 & -s-2 & -s-2 \\ 0 & -2s-4 & -2s-4 \end{bmatrix}$$

Hence, the structural graph of the system $\mathscr{S}$ is composed of two strongly connected subgraphs corresponding to vertex 1 (as the first subgraph), and vertices 2, 3 (as the second subgraph). Therefore, the new subsystem $\tilde{S}_1$ is defined to be the subsystem S_1, and $\tilde{S}_2$ is defined to be the union of S_2 and S_3. The union of the subsystems $\tilde{S}_1$ and $\tilde{S}_2$ is sometimes referred to as a quotient system [17]. It is important to note that:

- The DFMs of the system $\mathscr{S}(S_1, S_2, S_3)$ are also the modes of the closed-loop system in Figure 2.1, for any arbitrary dynamic LTI controllers K_1, K_2 and K_3. It can be easily verified that $\lambda = 1$ is the only DFM of the system $\mathscr{S}$ given by (2.3).
- The QFMs of the system $\mathscr{S}(S_1, S_2, S_3)$ are defined to be the DFMs of the system $\mathscr{S}(\tilde{S}_1, \tilde{S}_2)$, i.e., the fixed modes of the closed-loop system shown in Figure 2.2, for any arbitrary LTI controllers $\tilde{K}_1$ and $\tilde{K}_2$. For instance, it is easy to show that $\lambda = 1$ is a QFM of the system $\mathscr{S}$ given by (2.3).

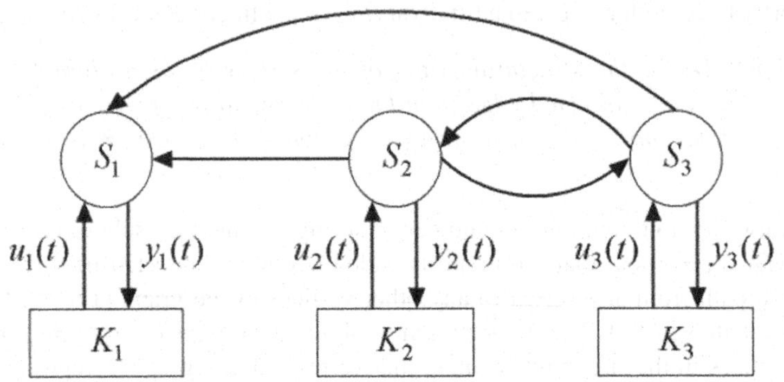

Fig. 2.1 The schematic of the decentralized control system $\mathscr{S}$ used for obtaining the DFMs.

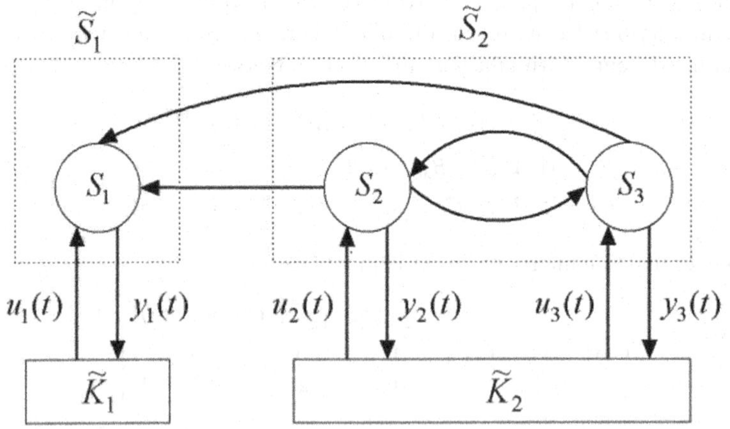

Fig. 2.2 The schematic of the decentralized control system $\mathscr{S}$ used for obtaining the QFMs.

2.3 Characterization of decentralized fixed modes

It is desired in this section to present a simple procedure to obtain the DFMs of the system $\mathscr{S}$.

Notation: For any $i, j \in \bar{\nu}$:

- Denote the (μ_1, μ_2) entry of B_i with $b_i^{\mu_1,\mu_2}$, for any $1 \leq \mu_1 \leq n$, $1 \leq \mu_2 \leq m_i$.
- Denote the (μ_1, μ_2) entry of C_i with $c_i^{\mu_1,\mu_2}$, for any $1 \leq \mu_1 \leq r_i$, $1 \leq \mu_2 \leq n$.
- Denote the (μ_1, μ_2) entry of D_{ij} with $d_{ij}^{\mu_1,\mu_2}$, for any $1 \leq \mu_1 \leq r_i$, $1 \leq \mu_2 \leq m_j$.

The following theorem formulates the DFMs of the system $\mathscr{S}$.

Theorem 1 *Assume that the mode σ_i, $i \in \{1,2,...,n\}$, is controllable as well as observable. σ_i is a DFM of the system $\mathscr{S}$, $\nu \geq 2$, if and only if there exist a permutation of $\{1,2,...,\nu\}$ denoted by distinct integers $i_1, i_2, ..., i_\nu$ and an integer $p \in [1, \nu-1]$ such that $b_\eta^{i,\alpha} = c_\gamma^{\beta,i} = 0$, and:*

$$\sum_{\mu=1,\ \mu\neq i}^{n} \frac{b_\eta^{\mu,\alpha} c_\gamma^{\beta,\mu}}{\sigma_\mu - \sigma_i} = d_{\gamma\eta}^{\beta,\alpha} \tag{2.4}$$

for all η, γ, α and β given by:

$$\eta \in \{i_1, i_2, ..., i_p\},\quad \gamma \in \{i_{p+1}, i_{p+2}, ..., i_\nu\},\quad 1 \leq \alpha \leq m_\eta,\quad 1 \leq \beta \leq r_\gamma \tag{2.5}$$

Proof: It is known that σ_i is a DFM of the system $\mathscr{S}(S_1, S_2, ..., S_\nu)$ if and only if there exist a permutation of $\{1,2,...,\nu\}$ denoted by distinct integers $i_1, i_2, ..., i_\nu$ and an integer $p \in [0, \nu]$ such that the rank of the following matrix is less than n [2]:

$$\begin{bmatrix} A - \sigma_i I_n & B_{i_1} & B_{i_2} & \cdots & B_{i_p} \\ C_{i_{p+1}} & D_{i_{p+1}i_1} & D_{i_{p+1}i_2} & \cdots & D_{i_{p+1}i_p} \\ C_{i_{p+2}} & D_{i_{p+2}i_1} & D_{i_{p+2}i_2} & \cdots & D_{i_{p+2}i_p} \\ \vdots & \vdots & \vdots & \ddots & \vdots \\ C_{i_\nu} & D_{i_\nu i_1} & D_{i_\nu i_2} & \cdots & D_{i_\nu i_p} \end{bmatrix} \tag{2.6}$$

In addition, since it is assumed that the mode σ_i is controllable and observable, the rank of the matrix (2.6) is equal to n for $p = 0$ and $p = \nu$. Therefore, the condition $0 \leq p \leq \nu$ given above can be reduced to $1 \leq p \leq \nu - 1$. It is clear that the rank of the matrix $A - \sigma_i I_n$ is $n-1$, and also, the ith column and the ith row of this matrix are both zeros. Hence, if there exists a nonzero entry either in the ith column or in the ith row of the matrix given in (2.6), its rank will be at least n. As a result, the rank of the matrix in (2.6) is less than n, if and only if both of the following conditions hold:

i) All of the entries of the ith column and the ith row of the matrix given in (2.6) are zero, i.e., $b_\eta^{i,\alpha} = c_\gamma^{\beta,i} = 0$ for any α, β, η, and γ satisfying (2.5).

ii) The rank of the following matrix:

$$\begin{bmatrix} \sigma_1^i & \dots & 0 & 0 & \dots & 0 & b_\eta^{1,\alpha} \\ \vdots & \ddots & \vdots & \vdots & \ddots & \vdots & \vdots \\ 0 & \dots & \sigma_{i-1}^i & 0 & \dots & 0 & b_\eta^{i-1,\alpha} \\ 0 & \dots & 0 & \sigma_{i+1}^i & \dots & 0 & b_\eta^{i+1,\alpha} \\ \vdots & \ddots & \vdots & \vdots & \ddots & \vdots & \vdots \\ 0 & \dots & 0 & 0 & \dots & \sigma_n^i & b_\eta^{n,\alpha} \\ c_\gamma^{\beta,1} & \dots & c_\gamma^{\beta,i-1} & c_\gamma^{\beta,i+1} & \dots & c_\gamma^{\beta,n} & d_{\gamma\eta}^{\beta,\alpha} \end{bmatrix} \tag{2.7}$$

(which is a sub-matrix of the one given by (2.6)) is less than n for any α, β, η, and γ satisfying (2.5), where $\sigma_j^i := \sigma_j - \sigma_i$, $i, j \in \{1, 2, ..., n\}$. Partition the matrix given by (2.7) into four sub-matrices, and denote it with $\begin{bmatrix} A_i & \Phi_1 \\ \Phi_2 & d_{\gamma\eta}^{\beta,\alpha} \end{bmatrix}$, where $A_i \in \mathfrak{R}^{(n-1)\times(n-1)}$, $\Phi_1 \in \mathfrak{R}^{(n-1)\times 1}$, and $\Phi_2 \in \mathfrak{R}^{1\times(n-1)}$. Since the matrix A_i is nonsingular (because it is assumed that $\sigma_1, ..., \sigma_n$ are distinct), one can write:

$$\det \begin{bmatrix} A_i & \Phi_1 \\ \Phi_2 & d_{\gamma\eta}^{\beta,\alpha} \end{bmatrix} = \det(A_i) \times \det\left(d_{\gamma\eta}^{\beta,\alpha} - \Phi_2 A_i^{-1} \Phi_1\right)$$

Thus, the rank of the matrix given in (2.7) is less than n, if and only if the scalar $\Phi_2 A_i^{-1} \Phi_1$ is equal to $d_{\gamma\eta}^{\beta,\alpha}$, i.e.:

$$\sum_{\mu=1,\ \mu\neq i}^{n} \frac{b_\eta^{\mu,\alpha} c_\gamma^{\beta,\mu}}{\sigma_\mu^i} = d_{\gamma\eta}^{\beta,\alpha} \tag{2.8}$$

■

Define now the matrix M_i as:

$$M_i := C \times \mathrm{diag}\left(\left[\frac{1}{\sigma_1 - \sigma_i}, \dots, \frac{1}{\sigma_{i-1} - \sigma_i}, 0, \frac{1}{\sigma_{i+1} - \sigma_i}, \dots, \frac{1}{\sigma_\nu - \sigma_i}\right]\right) B - D$$

and denote its (μ_1, μ_2) block entry with $M_i^{\mu_1,\mu_2} \in \mathfrak{R}^{r_{\mu_1} \times m_{\mu_2}}$, for any $\mu_1, \mu_2 \in \bar{\nu}$.

Theorem 2 *The mode σ_i, $i \in \{1, 2, ..., n\}$, is a DFM of the system $\mathscr{S}$, $\nu \geq 2$, if and only if any of the following conditions holds:*

(i) The ith row of the matrices $B_1, B_2, ..., B_\nu$ are zero.
ii) The ith column of the matrices $C_1, C_2, ..., C_\nu$ are zero.
iii)There exist a permutation of $\{1, 2, ..., \nu\}$ denoted by distinct integers $i_1, i_2, ..., i_\nu$ and an integer $p \in [1, \nu - 1]$ such that $M_i^{\gamma,\eta}$ is a zero matrix for any $\eta \in \{i_1, i_2, ..., i_p\}$ and $\gamma \in \{i_{p+1}, i_{p+2}, ..., i_\nu\}$, and moreover the ith row of the matrices $B_1, B_2, ..., B_{i_p}$ and the ith column of $C_{i_{p+1}}, C_{i_{p+2}}, ..., C_{i_\nu}$ are all zero.

Proof: Criteria (i) and (ii) are equivalent to the uncontrollability and the unobservability, respectively. Furthermore, Criterion (iii) is resulted from Theorem 1, on noting that $M_i^{\gamma,\eta}$ is a $r_\gamma \times m_\eta$ matrix whose (β,α) entry is equal to:

$$\sum_{\mu=1,\ \mu\neq i}^{n} \frac{b_\eta^{\mu,\alpha} c_\gamma^{\beta,\mu}}{\sigma_\mu - \sigma_i} - d_{\gamma\eta}^{\beta,\alpha} \tag{2.9}$$

for any integers $\beta \in [1, r_\gamma]$, $\alpha \in [1, m_\eta]$. ■

It is desired now to construct a graph based on the matrix M_i. Consider a bipartite graph $\mathscr{G}_i$ with ν vertices $1, 2, ..., \nu$ in each of its vertex sets, namely set 1 and set 2. For any $\mu_1, \mu_2 \in \bar{\nu}$, connect vertex μ_1 of set 1 to vertex μ_2 of set 2 if the matrix $M_i^{\mu_1,\mu_2}$ is a zero matrix. Then, mark vertex μ_1 of set 1 if the ith column of the matrix C_{μ_1} is a zero vector, for any $\mu_1 \in \bar{\nu}$. Likewise, mark vertex μ_2 of set 2 if the ith row of the matrix B_{μ_2} is a zero vector, for any $\mu_2 \in \bar{\nu}$.

The following algorithm results from Theorem 2 for verifying whether or not the mode σ_i is a DFM of the system $\mathscr{S}$.

Algorithm 1:

Step 1) Compute the matrix M_i, and construct the graph $\mathscr{G}_i$ in terms of it, as pointed out earlier.
Step 2) Verify if all of the vertices in set 1 of the graph $\mathscr{G}_i$ are marked. If yes, go to Step 6.
Step 3) Verify if all of the vertices in set 2 of the graph $\mathscr{G}_i$ are marked. If yes, go to Step 6.
Step 4) Check whether the graph $\mathscr{G}_i$ includes a complete bipartite subgraph such that all of its vertices are marked and moreover the set of the indices of its vertices is equal to the set $\bar{\nu}$. If yes, go to Step 6.
Step 5) The mode σ_i is not a DFM of the system $\mathscr{S}$. Stop the algorithm.
Step 6) The mode σ_i is a DFM of the system $\mathscr{S}$. Stop the algorithm.

Algorithm 1 proposes a simple graph-theoretic approach to find the DFMs of the system $\mathscr{S}$. This method requires deriving a certain matrix, and then checking the existence of a complete subgraph in a graph, which can be accomplished using numerous efficient algorithms. In contrast, the existing methods require the rank of several matrices (say 2^ν) to be checked, which can be cumbersome when the matrix is of high dimension. In fact, the above algorithm presents a simple combinatorial procedure as a more efficient alternative to find the DFMs of a system (with distinct modes).

Corollary 1 *Denote the number of matrices $B_1, B_2, ..., B_\nu$ whose i^{th} row are zero with Γ_i. Furthermore, denote the number of matrices $C_1, C_2, ..., C_\nu$ whose i^{th} column are zero with $\bar{\Gamma}_i$. If $\Gamma_i + \bar{\Gamma}_i$ is less than ν, then σ_i is not a DFM of the system $\mathscr{S}$.*

Proof: It is straightforward to show that if $\Gamma_i + \bar{\Gamma}_i$ is less than ν, none of Steps 2, 3 or 4 of Algorithm 1 is fulfilled. ■

Corollary 1 presents a quite simple test as a sufficient condition to verify whether σ_i can be a DFM of the system or not.

2.4 Characterization of quotient fixed modes

It is desired now to present a graph-theoretic approach to obtain the QFMs of the system $\mathscr{S}$, similar to the one introduced for the DFMs in the preceding section. Since QFM is merely defined for the strictly proper systems, it will be assumed hereafter that $D = 0$.

Theorem 3 *The mode σ_i is a QFM of the system $\mathscr{S}$, $\nu \geq 2$ if and only if either condition (a) or condition (b) given below holds:*

a) σ_i is an uncontrollable or unobservable mode.
b) There exist a permutation of $\{1,2,...,\nu\}$ denoted by distinct integers $i_1, i_2, ..., i_\nu$ and an integer $p \in [1, \nu-1]$ such that for all η and γ given by:

$$\eta \in \{i_1, i_2, ..., i_p\}, \quad \gamma \in \{i_{p+1}, i_{p+2}, ..., i_\nu\} \tag{2.10}$$

both of the conditions given below hold:

i) The ith column of the matrix C_γ and the ith row of the matrix B_η are both zero vectors.
ii) Consider the jth column of the matrix C_γ and the jth row of the matrix B_η; at least one of these two vectors is zero, for any $j \in \{1,2,...,n\}$.

Proof: It is trivial to show that if condition (a) in Theorem 3 holds, the mode σ_i will be a QFM of the system $\mathscr{S}$. If it does not hold, then it follows directly from Theorems 1 and 2 given in [3], that σ_i is a QFM if and only if there exist a permutation of $\{1,2,...,\nu\}$ denoted by distinct integers $i_1, i_2, ..., i_\nu$ and an integer $p \in [1, \nu-1]$ such that $b_\eta^{i,\alpha} = c_\gamma^{\beta,i} = 0$, and $b_\eta^{\mu,\alpha} c_\gamma^{\beta,\mu} = 0$ for all η, γ, α and β given by (2.5) and $\mu \in \{1,2...,n\}$. It is straightforward to show that this requirement is identical to condition (b) in the theorem. ■

Consider a bipartite graph $\bar{\mathscr{G}}_i$ with ν vertices $1,2,...,\nu$ in each of its vertex sets, namely set 1 and set 2. For any $\mu_1, \mu_2 \in \bar{\nu}$, connect vertex μ_1 of set 1 to vertex μ_2 of set 2 if either the jth column of C_{μ_1} or the jth row of B_{μ_2} is a zero vector for all $j \in \{1,2,...,n\}$. Then, mark vertex μ_1 of set 1 if the ith column of the matrix C_{μ_1} is a zero vector, for any $\mu_1 \in \bar{\nu}$. Likewise, mark vertex μ_2 of set 2 if the ith row of the matrix B_{μ_2} is a zero vector, for any $\mu_2 \in \bar{\nu}$. It is worth noting that the graphs $\bar{\mathscr{G}}_1, \bar{\mathscr{G}}_2, ..., \bar{\mathscr{G}}_\nu$ have the same edges, although the marking of their vertices might be different.

The following algorithm results from Theorem 3 for verifying whether or not the mode σ_i is a QFM of the system $\mathscr{S}$.

Algorithm 2:

Step 1) Construct the graph $\bar{\mathscr{G}}_i$, as discussed above.
Step 2) Verify if all of the vertices in set 1 of the graph $\bar{\mathscr{G}}_i$ are marked. If yes, go to Step 6.
Step 3) Verify if all of the vertices in set 2 of the graph $\bar{\mathscr{G}}_i$ are marked. If yes, go to Step 6.

Step 4) Check whether the graph $\bar{\mathscr{G}}_i$ includes a complete bipartite subgraph such that all of its vertices are marked and the set of the indices of its vertices is equal to the set $\bar{\nu}$. If yes, go to Step 6.
Step 5) The mode σ_i is not a QFM of the system $\mathscr{S}$. Stop the algorithm.
Step 6) The mode σ_i is a QFM of the system $\mathscr{S}$. Stop the algorithm.

2.5 Illustrative examples

Example 1: Consider a system $\mathscr{S}$ consisting of five single-input single-output (SISO) subsystems with the following state-space matrices:

$$A = \text{diag}([1\ ,\ 2\ ,\ 3\ ,\ 4\ ,\ 5]), \quad B = \begin{bmatrix} 0 & 0 & 0 & 1 & 2 \\ 2 & 1 & 3 & 1 & 4 \\ 0 & 2 & 4 & -1 & 5 \\ 0 & 0 & 3 & 0 & -3 \\ 0 & 0 & 0 & 3 & -1 \end{bmatrix},$$
$$C = \begin{bmatrix} 0 & 3 & 2 & 1 & 4 \\ 0 & 3 & 4 & 2 & -1 \\ 5 & 4 & 3 & -2 & 4 \\ 0 & 2 & 3 & 1 & 3 \\ 0 & -2 & -3 & -2 & -4 \end{bmatrix}, \quad D = \begin{bmatrix} 6 & 5 & 14 & 3 & 2 \\ 6 & 7 & 19 & 4 & 2 \\ 8 & 7 & 16 & -2 & -4 \\ 4 & 5 & 13 & 0 & 1 \\ -4 & -5 & -14 & -1 & 2 \end{bmatrix} \tag{2.11}$$

It is desired to verify which of the modes $\sigma_i = i$, $i \in \bar{\nu} = \{1,2,3,4,5\}$, are DFMs of the system $\mathscr{S}$. First, let the test given in Corollary 1 be carried out. Since the first entries of $B_1, B_2, B_3, C_1, C_2, C_4$ and C_5 are all zero, $\Gamma_1 + \bar{\Gamma}_1$ is equal to 7. Similarly, one can conclude that:

$$\Gamma_2 + \bar{\Gamma}_2 = 0, \;\; \Gamma_3 + \bar{\Gamma}_3 = 1, \;\; \Gamma_4 + \bar{\Gamma}_4 = 3, \;\; \Gamma_5 + \bar{\Gamma}_5 = 3 \tag{2.12}$$

Due to the fact that $\Gamma_i + \bar{\Gamma}_i < 5$ for $i = 2,3,4,5$, it follows from Corollary 1 that none of the modes $2,3,4$ and 5 is a DFM of the system $\mathscr{S}$. Algorithm 1 will now be used to find out whether $\sigma_1 = 1$ is a DFM. The matrix M_1 will be obtained as:

$$\begin{aligned} M_1 &= C \times \text{diag}\left(\left[0\ ,\ 1\ ,\ \frac{1}{2}\ ,\ \frac{1}{3}\ ,\ \frac{1}{4}\right]\right) B - D \\ &= \begin{bmatrix} 0 & 0 & 0 & 2 & 13 \\ 0 & 0 & 0 & -3.75 & 18.25 \\ 0 & 0 & 0 & 7.5 & 28.5 \\ 0 & 0 & 0 & 2.75 & 12.75 \\ 0 & 0 & 0 & -2.5 & -14.5 \end{bmatrix} \end{aligned} \tag{2.13}$$

The graph $\mathscr{G}_1$ corresponding to the matrix M_1 is sketched in Figure 2.3. Since the first entries of $B_1, B_2, B_3, C_1, C_2, C_4$ and C_5 are all zero, vertices 1, 2 and 3 from set 2, and vertices 1, 2, 4 and 5 from set 1 of the graph $\mathscr{G}_1$ are marked by filled circles, as shown in the figure. It can be easily observed that vertices 4, 5 of set 1 and vertices 1, 2, 3 of set 2 fulfill the following criteria:

- All of them are marked.
- They constitute a complete bipartite graph.
- The set of their labels is equal to $\bar{\mathcal{V}}$.

Therefore, $\sigma_1 = 1$ is a DFM of the system (from Step 4 of Algorithm 1).

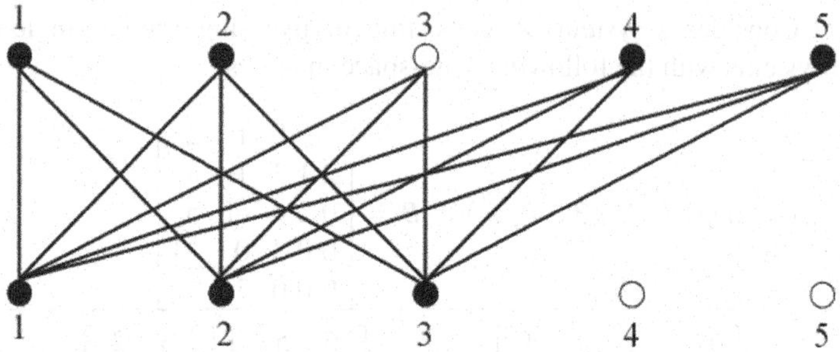

Fig. 2.3 The graph $\mathscr{G}_1$ corresponding to the matrix M_1 given in (2.13).

Regarding the mode $\sigma_3 = 3$, let Algorithm 1 be pursued for this mode regardless of the observation that it failed the test given in Corollary 1. The matrix M_3 is equal to:

$$
\begin{aligned}
M_3 &= C \times \operatorname{diag}\left(\left[-1\,,\,0\,,\,1\,,\,\frac{1}{2}\,,\,\frac{1}{3}\right]\right) B - D \\
&= \begin{bmatrix} -12 & -8 & -20 & 0 & -19 \\ -12 & -10 & -22 & -8.5 & -19 \\ -16 & -11 & -34 & 1.5 & -13 \\ -8 & -7 & -16 & 2.5 & -13 \\ 8 & 7 & 14 & -3 & 14 \end{bmatrix}
\end{aligned}
\tag{2.14}
$$

The corresponding graph $\mathscr{G}_3$ is depicted in Figure 2.4. Since there are not enough edges in the graph to create a complete bipartite subgraph which spans all the indices, thus $\sigma_3 = 3$ is not a DFM of the system (which also confirms the result obtained from Corollary 1).

Consequently, the system has only one DFM at 1. This result could also be obtained by using the method given in [2] or [15], which require the rank of 5×2^5 matrices with the dimensions between 5 and 10 be checked. The sizable difference between the computational requirements of the method presented in this chapter and the ones given in [2, 15] demonstrates the efficacy of this work. It is worth mentioning that the results obtained here by using the proposed method are attained by hand, while the methods given in [2, 15] require a proper software (such as MATLAB)

Example 2: Consider a strictly proper system $\mathscr{S}$ consisting of three two-input two-output subsystems with the following parameters:

$$A = \text{diag}([1\ ,\ 2\ ,\ 3\ ,\ 4\ ,\ 5]),$$
$$B_1 = \begin{bmatrix} 0 & 0 \\ 0 & 0 \\ 0 & 0 \\ 2 & -1 \\ 0 & 0 \end{bmatrix},\quad B_2 = \begin{bmatrix} 0 & 0 \\ 2 & 1 \\ 0 & 0 \\ 1 & 2 \\ 0 & 0 \end{bmatrix},\quad B_3 = \begin{bmatrix} 1 & 1 \\ -1 & -1 \\ 1 & -1 \\ -1 & 1 \\ -1 & -1 \end{bmatrix},$$
$$C_1 = \begin{bmatrix} 1 & -1 \\ -1 & 1 \\ 1 & -1 \\ -1 & 1 \\ 1 & -1 \end{bmatrix}^T,\quad C_2 = \begin{bmatrix} 2 & 3 \\ 0 & 0 \\ 0 & 0 \\ 0 & 0 \\ 1 & 1 \end{bmatrix}^T,\quad C_3 = \begin{bmatrix} 5 & 6 \\ 0 & 0 \\ 0 & 0 \\ 0 & 0 \\ -1 & -1 \end{bmatrix}^T \tag{2.15}$$

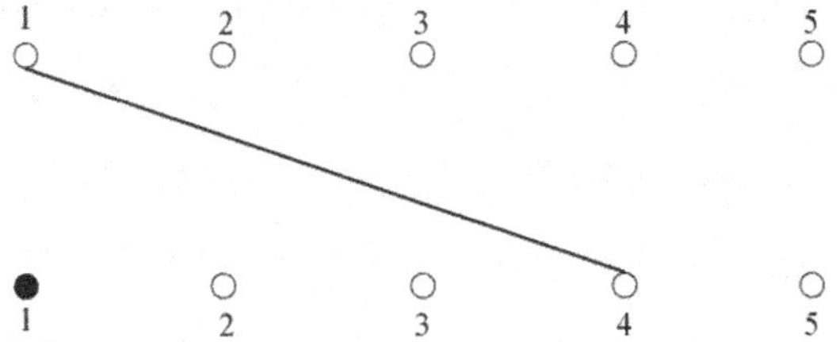

Fig. 2.4 The graph $\mathscr{G}_3$ corresponding to the matrix M_3 given in (2.14).

The graphs $\bar{\mathscr{G}}_i$, $i \in \{1,2,3,4,5\}$ are depicted in Figures 2.5, 2.6 and 2.7. Using Algorithm 2, it can be concluded that $\sigma_2 = 2$ is a QFM of the system, as Step 4 will be satisfied by considering vertices 2 and 3 from set 1, and vertex 1 from set 2. Likewise, the mode $\sigma_3 = 3$ is a QFM by considering either *vertices 2 and 3 from set 1, and vertex 1 from set 2* or *vertex 3 from set 1, and vertices 1 and 2 from set* 2. It can be easily verified that none of the remaining modes are QFMs of the system $\mathscr{S}$.

2.6 Summary

This chapter aims to characterize the fixed modes of a decentralized system with distinct modes. First, decentralized fixed modes (DFM) are described using graph-theoretic techniques. Then, quotient fixed modes (QFM), which are immovable with respect to any type of decentralized control law, are characterized. Unlike the existing methods which require the computation of the rank of several matrices, the approaches proposed here transform the knowledge of the system into bipartite graphs.

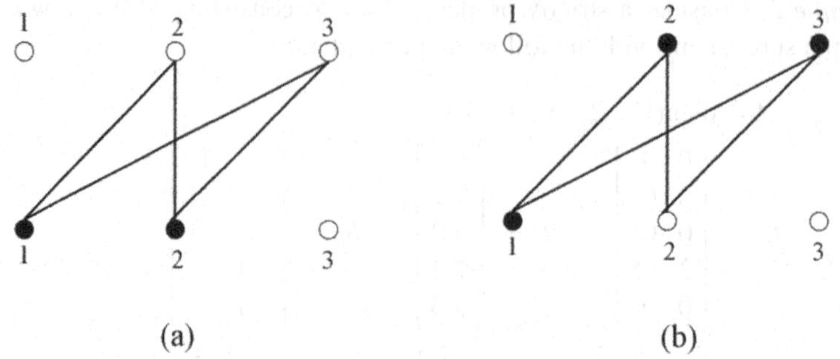

Fig. 2.5 The graphs $\bar{\mathscr{G}}_1$ and $\bar{\mathscr{G}}_2$ are sketched in (a) and (b), respectively.

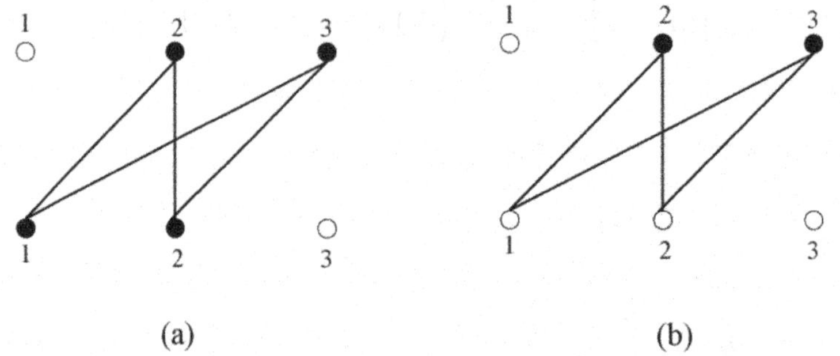

Fig. 2.6 The graphs $\bar{\mathscr{G}}_3$ and $\bar{\mathscr{G}}_4$ are sketched in (a) and (b), respectively.

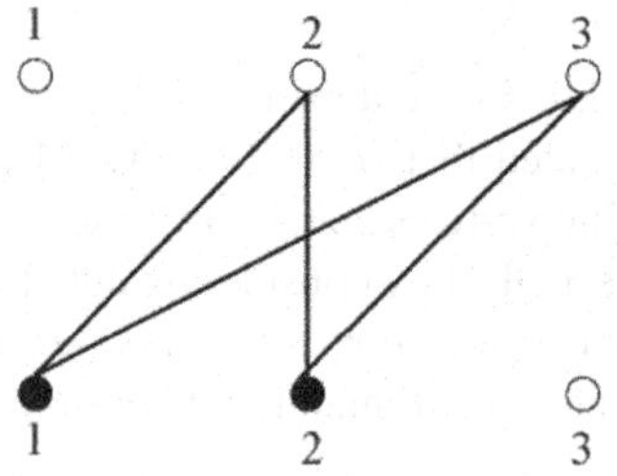

Fig. 2.7 The graphs $\bar{\mathscr{G}}_5$.

Then, it is asserted that finding a complete bipartite subgraph with a certain property is equivalent to the existence of a DFM. A similar result is attained for the QFMs. The efficacy of the proposed method is demonstrated through numerical examples.

References

1. S. H. Wang and E. J. Davison, "On the stabilization of decentralized control systems," *IEEE Transactions on Automatic Control*, vol. 18, no. 5, pp. 473-478, 1973.
2. E. J. Davison and T. N. Chang, "Decentralized stabilization and pole assignment for general proper systems," *IEEE Transactions on Automatic Control*, vol. 35, no. 6, pp. 652-664, 1990.
3. J. Lavaei and A. G. Aghdam, "Elimination of fixed modes by means of high-performance constrained periodic control," in *Proceedings of 45th IEEE Conference on Decision and Control*, San Diego, CA, pp. 4441-4447, 2006.
4. D. D. Šiljak, Decentralized control of complex systems, Cambridge: Academic Press, 1991.
5. D. D. Šiljak and A. I. Zecevic, "Control of large-scale systems: Beyond decentralized feedback," *Annual Reviews in Control*, vol. 29, no. 2, pp. 169-179, 2005.
6. J. Lavaei and A. G. Aghdam, "A necessary and sufficient condition for the existence of a LTI stabilizing decentralized overlapping controller," in *Proceedings of 45th IEEE Conference on Decision and Control*, San Diego, CA, pp. 6179-6186, 2006.
7. G. Inalhan, D. M. Stipanovic, and C. J. Tomlin, "Decentralized optimization with application to multiple aircraft coordination," in *Proceedings of 41st IEEE Conference on Decision and Control*, Vegas, NV., pp. 1147-1155, 2002.
8. J. Lavaei, A. Momeni and A. G. Aghdam, "High-performance decentralized control for formation flying with leader-follower structure," in *Proceedings of 45th IEEE Conference on Decision and Control*, San Diego, CA, pp. 5947-5954, 2006.
9. J. Lavaei and A. G. Aghdam, "Decentralized control design for interconnected systems based on a centralized reference controller," in *Proceedings of 45th IEEE Conference on Decision and Control*, San Diego, CA, pp. 1189-1195, 2006.
10. S. S. Stankovic, M. J. Stanojevic, and D. D. Šiljak, "Decentralized overlapping control of a platoon of vehicles," *IEEE Transactions on Control Systems Technology*, vol. 8, no. 5, pp. 816-832, 2000.
11. J. Lavaei and A. G. Aghdam, "Optimal periodic feedback design for continuous-time LTI systems with constrained control structure," *International Journal of Control*, vol. 80, no. 2, pp. 220-230, 2007.
12. J. Lavaei and A. G. Aghdam, "Simultaneous LQ control of a set of LTI systems using constrained generalized sampled-data hold functions," *Automatica*, vol. 43, no. 2, pp. 274-280, 2007.
13. B. L. O. Anderson and D. J. Clements, "Algebraic characterizations of fixed modes in decentralized systems," *Automatica*, vol. 17, no. 5, pp. 703-712, 1981.
14. B. L. O. Anderson, "Transfer function matrix description of decentralized fixed modes," *IEEE Transactions on Automatic Control*, vol. 27, no. 6, pp. 1176-1182, 1982.
15. E. J. Davison and S. H. Wang, "A characterization of decentralized fixed modes in terms of transmission zeros," *IEEE Transactions on Automatic Control*, vol. 30, no. 1, pp. 81-82, 1985.
16. Z. Gong and M. Aldeen, "On the characterization of fixed modes in decentralized control," *IEEE Transactions on Automatic Control*, vol. 37, no. 7, pp. 1046-1050, 1992.
17. Z. Gong and M. Aldeen, "Stabilization of decentralized control systems," *Journal of Mathematical Systems, Estimation, and Control*, vol. 7, no. 1, pp. 1-16, 1997.

Chapter 3
Time Complexity of Decentralized Fixed Mode Verification

3.1 Introduction

An interconnected system consists of a number of interacting subsystems, which could be homogeneous or heterogeneous. It is evident that many real-world systems can be modeled as interconnected systems, some of which are communication networks, large space structures, power systems, and chemical processes [1, 2, 3, 4, 5]. The classical control techniques often fail to control such systems, in light of some well-known practical issues such as computation or communication constraints. This has given rise to the emergence of the decentralized control area that aims to design non-classical structurally constrained controllers [6]. More precisely, a (conventional) decentralized controller comprises a set of non-interacting local controllers corresponding to different subsystems.

The notion of a decentralized fixed modes (DFM) was introduced in [7] to characterize those modes of the system which cannot be moved using a linear time-invariant (LTI) decentralized controller. Several methods have been proposed in the literature to find the DFMs of a system [8, 9, 10, 11]. For instance, an algebraic characterization of DFMs was provided in [8]. The method given in [9], on the other hand, characterizes the DFMs of a system in terms of its transfer function. It was also shown in [12] that the DFMs of any system can be found by checking the transmission zeros of a set of artificial systems derived from the original system. An algorithm was presented in [10] to identify the DFMs of a system by computing the rank of a set of matrices. Unfortunately, the number of systems whose transmission zeros need to be checked in [12] and the number of matrices whose ranks are to be computed in [10] depend exponentially on the number of subsystems of the original system.

On the other hand, a method is delineated in [13] stating that in order to numerically find the DFMs of a system, it is sufficient to apply a randomly generated static decentralized controller to the system, and then verify what modes of the system are still fixed. Nevertheless, this method is often inaccurate for large-scale systems. More precisely, calculating the eigenvalues of a large-size matrix is normally associated with some errors (especially when the matrix has complex eigenvalues), which makes it impossible to distinguish the fixed modes from the approximate fixed modes [14]. Another issue is that a generic static decentralized controller may not be able to sufficiently displace a mode so that it is recognized as not being a DFM.

A graph-theoretic method was proposed in the recent work [11], which constructs a bipartite graph corresponding to each unrepeated mode of the system. This work states that the mode is a DFM if and only if the graph contains a bipartite subgraph satisfying two specific properties. Although this method seems to be far simpler than the other available deterministic methods, it is not clear how to systematically verify the existence of such a subgraph.

Consider a decision problem, whose answer to be found is "yes" or "no". An algorithm provided for this problem is said to be *efficient* if its time complexity is satisfactory. Informally speaking, the time complexity measures the number of machine instructions executed during the running time of the algorithm (as a function of the size of the input). It is well-understood in computer science that if there exists an efficient randomized algorithm for a decision problem, the existence of a deterministic algorithm with a similar complexity is expected. In other words, randomized algorithms cannot be far more efficient than deterministic algorithms. Regarding the decentralized question posed here (i.e. finding the DFMs of a system), the work [13] shows that there exists an efficient randomized algorithm, whereas the available deterministic algorithms have high time complexities. Based on the above-mentioned discussion, one would conjecture that there exists an efficient deterministic algorithm for the underlying decentralized problem. Finding such an algorithm and investigating its properties are central to the current work.

Given an LTI interconnected system realized in a canonical form, consider a distinct mode of the system. The primary objective of this chapter is to determine the time complexity of deciding whether this mode is a DFM of the system. To tackle this decision problem, a digraph is constructed by means of an algorithm, whose time complexity is the same as the complexity of matrix multiplication. It is then shown that the answer to the posed decision problem is affirmative if and only if the digraph is not strongly connected. The time complexity of the latter problem (i.e., checking the strong connectivity) is quadratic with respect to the number of subsystems of the system. It is eventually concluded that the time complexity of the original decision problem is the same as that of matrix multiplication, being at most equal to $O(n^{2.376})$, where n denotes the order of the given system. Note that it is unlikely to find another algorithm for this decentralized problem whose complexity is less than the one provided in the present work, due to the fact that any possible algorithm is not permitted to use ordinary matrix operations such as *multiplication, inversion, determinant, rank,* etc. over arbitrary unstructured matrices. This signifies that the obtained complexity order is believed to be the best possible one.

3.2 Preliminaries and problem formulation

Consider an LTI interconnected system $\mathscr{S}$ consisting of ν subsystems $S_1, S_2, ..., S_\nu$, represented by

$$\begin{aligned} \dot{x}(t) &= Ax(t) + \sum_{j=1}^{\nu} B_j u_j(t) \\ y_i(t) &= C_i x(t) + \sum_{j=1}^{\nu} D_{ij} u_j(t), \quad i \in \boldsymbol{\nu} := \{1, 2, ..., \nu\}, \end{aligned} \tag{3.1}$$

where $x(t) \in \Re^n$ is the state, and $u_i(t) \in \Re^{m_i}$ and $y_i(t) \in \Re^{r_i}$, $i \in \boldsymbol{\nu}$, are the input and output of the ith subsystem, respectively. Define

$$\begin{aligned} B &:= \begin{bmatrix} B_1 & \cdots & B_\nu \end{bmatrix}, \quad C := \begin{bmatrix} C_1 \\ \vdots \\ C_\nu \end{bmatrix}, \\ D &:= \begin{bmatrix} D_{11} & \cdots & D_{1\nu} \\ \vdots & \ddots & \vdots \\ D_{\nu 1} & \cdots & D_{\nu\nu} \end{bmatrix}, \\ m &:= \sum_{i=1}^{\nu} m_i, \quad r := \sum_{i=1}^{\nu} r_i. \end{aligned} \tag{3.2}$$

A (conventional) decentralized controller for the system $\mathscr{S}$ is composed of a set of ν local controllers, where the ith local controller, $i \in \boldsymbol{\nu}$, observes only the local output $y_i(t)$ to construct the local input $u_i(t)$ of the ith subsystem. The following definition was presented in [7] for strictly proper systems and generalized in [10] for proper systems.

Definition 1 *A mode σ is said to be a decentralized fixed mode (DFM) of the system $\mathscr{S}$ if there exists no static decentralized controller to displace this mode. In other words, σ is a DFM of the system $\mathscr{S}$ if the relation*

$$\sigma \in sp\left(A + BK(I - DK)^{-1}C\right) \tag{3.3}$$

holds for every block diagonal matrix K whose i^{th} block entry, $i \in \boldsymbol{\nu}$, is a matrix of dimension $m_i \times r_i$, where $sp(\cdot)$ stands for the matrix spectrum.

It is noteworthy that as shown in [10], a DFM is fixed with respect to not only static decentralized controllers but also all types of LTI decentralized controllers. In what follows, different methods for finding the DFMs of a system are outlined. Note that for a better understanding of this work, some useful concepts in graph theory are presented in an appendix given at the end of the chapter.

3.2.1 Matrix rank checking

According to [10], a (centralized) controllable and observable mode σ is a DFM of the system $\mathscr{S}$ if and only if there exist a permutation of $\{1,2,...,\nu\}$ denoted by $(i_1,i_2,...,i_\nu)$ and an integer $p \in \{1,2,...,\nu-1\}$ such that the rank of the following matrix is less than n:

$$\begin{bmatrix} A-\sigma I_n & B_{i_1} & B_{i_2} & \dots & B_{i_p} \\ C_{i_{p+1}} & D_{i_{p+1}i_1} & D_{i_{p+1}i_2} & \dots & D_{i_{p+1}i_p} \\ C_{i_{p+2}} & D_{i_{p+2}i_1} & D_{i_{p+2}i_2} & \dots & D_{i_{p+2}i_p} \\ \vdots & \vdots & \vdots & \ddots & \vdots \\ C_{i_\nu} & D_{i_\nu i_1} & D_{i_\nu i_2} & \dots & D_{i_\nu i_p} \end{bmatrix}. \tag{3.4}$$

The number of matrices in the above form is equal to $(\nu-1)\times\nu!$. However, it is not required to compute the rank of all such matrices, as those ones which can be converted to each other via re-ordering rows and columns have the same rank. In fact, one needs to find the rank of only $2^\nu-2$ matrices, which is still an exponential number. This clearly signifies that computing the DFMs of the system $\mathscr{S}$ using this method is cumbersome.

3.2.2 Randomized algorithm

Pick a matrix $K \in \Re^{r\times m}$ at random and consider the matrices A and $A+BK(I-DK)^{-1}C$. The works [13] and [10] state that the DFMs of the system are, almost surely, the common eigenvalues of these two matrices. This gives rise to a randomized algorithm that almost always works correctly. As explained in the introduction, this method suffers from some numerical issues. Nonetheless, this technique indicates that there exists a simple randomized algorithm for finding DFMs, whose complexity is much lower than the deterministic one explained earlier (i.e., testing the ranks of an exponential number of matrices).

3.2.3 Derandomization

The work [10] proposes a derandomization technique for the randomized algorithm given in [13] (discussed above). Observe that a decentralized gain matrix K has $\sum_{i=1}^{\nu} m_i r_i$ free parameters, sitting on the block diagonal of this matrix. For every nonnegative integer j satisfying the inequality $j \le \sum_{i=1}^{\nu} m_i r_i$, pick j of these free parameters, give arbitrary nonzero values to them, and set the remaining free parameters to zero. This leads to a structured decentralized gain matrix. Repeating this procedure for all possible combinations yields p block-diagonal matrices $K_1,K_2,...,K_p$, where

$$p = 2^{\sum_{i=1}^{\nu} m_i r_i}. \tag{3.5}$$

The derandomized algorithm says that the DFMs of the system are the common eigenvalues of the matrices $A + BK_i(I - DK_i)^{-1}C,\ i = 1,2,...,p$. Note that even though this algorithm requires that some elements of K_i be given *arbitrary* nonzero values, this method is not really a randomized algorithm. The reason is that those elements need not be chosen using a pseudo-random number generator, and can all be simply considered equal to 1. Observe that although the randomized algorithm mentioned in the preceding subsection runs in polynomial time, its derandomized counterpart runs in exponential time.

3.2.4 Graph-theoretic approach

Assume that σ is an eigenvalue of A with multiplicity 1, which is also a (centralized) observable and controllable mode of the system $\mathscr{S}$. With no loss of generality, suppose that the matrix A is in the canonical form

$$A = \begin{bmatrix} \sigma & 0 \\ 0 & \mathbf{A} \end{bmatrix}, \tag{3.6}$$

where $\mathbf{A}$ is a matrix of appropriate dimension (this can be achieved using a proper similarity transformation, if need be). Define

$$M(\sigma) := \mathbf{C}(\sigma I_{n-1} - \mathbf{A})^{-1}\mathbf{B} + D, \tag{3.7}$$

where:

- $\mathbf{C}$ is derived from C by eliminating its first column.
- $\mathbf{B}$ is obtained from B by removing its first row.

Denote the (i, j) block entry of $M(\sigma)$ with $M_{ij}(\sigma) \in \Re^{r_i \times m_j}$, for every $i, j \in \nu$.

Definition 2 *Let $\mathscr{G}(\sigma)$ be a bipartite graph constructed as follows:*

- *Consider two sets of vertices, namely set 1 and set 2, with ν vertices in each of them.*
- *For every $i, j \in \nu$, $i \neq j$, connect vertex i in set 1 to vertex j in set 2 if all of the following conditions are satisfied:*
  - *The first column of C_i is zero.*
  - *The first row of B_j is zero.*
  - *$M_{ij}(\sigma)$ is a zero matrix*

We proposed the next result in [11] to verify whether or not σ is a DFM of the system $\mathscr{S}$.

Theorem 1 *The mode σ is a DFM of the system $\mathscr{S}$ if and only if the graph $\mathscr{G}(\sigma)$ contains a subgraph $\mathscr{G}_o(\sigma)$ with the following properties:*

- *It is complete bipartite.*
- *If $(i_1, i_2, ..., i_p)$ and $(j_1, j_2, ..., j_q)$ represent the sets of vertices of $\mathscr{G}_o(\sigma)$ (i.e. set 1 and set 2 of $\mathscr{G}_o(\sigma)$), then $(i_1, ..., i_p, j_1, ..., j_q)$ is a permutation of the set $\mathbf{v}$.*

Although this method seems to be far simpler than the deterministic methods outlined above, it is not clear how to verify the existence of such a subgraph $\mathscr{G}_o(\sigma)$ systematically.

3.2.5 *Problem statement*

This chapter develops the result of [11] under the assumption that the matrix A is in the canonical form (3.6). The objective is twofold. First, it is desired to propose a simple deterministic algorithm to check whether σ is a DFM of the system $\mathscr{S}$. Second, it is aimed to study the time complexity of this decision problem using deterministic algorithms.

3.3 Main results

Assume that the quantities m, r, ν are all less than or equal to n. This realistic assumption is made so that the complexity of computing the DFMs of $\mathscr{S}$ can be written only in terms of n. The following definitions turn out to be convenient in proceeding with the development of the chapter.

Definition 3 *Define $\tilde{\mathscr{G}}(\sigma)$ to be a directed graph (digraph) constructed as follows:*

- *Consider ν vertices, labeled as $1, 2, ..., \nu$.*
- *For every $i, j \in \mathbf{v}$, $i \neq j$, connect vertex i to vertex j by means of a directed edge if any of the conditions given below is satisfied:*
  - *The first column of C_i is a nonzero vector.*
  - *The first row of B_j is a nonzero vector.*
  - *$M_{ij}(\sigma)$ is not a zero matrix.*

It is well-known from graph theory that $\tilde{\mathscr{G}}(\sigma)$ can be uniquely decomposed as a union of strongly connected components, namely $\mathscr{C}_1, \mathscr{C}_2, ..., \mathscr{C}_k$, such that:

- $\mathscr{C}_i$, $i = 1, 2, ..., k$, is a strongly connected induced subgraph of $\tilde{\mathscr{G}}(\sigma)$.
- For every $i, j \in \{1, 2, ..., k\}$, $i < j$, there is no directed edge going from $\mathscr{C}_i$ to $\mathscr{C}_j$.

This fact will be exploited in the sequel to present one of the main results of this chapter.

Theorem 2 *The mode σ is a DFM of the system $\mathscr{S}$ if and only if the digraph $\tilde{\mathscr{G}}(\sigma)$ is not strongly connected.*

Proof of sufficiency: Assume that the digraph $\tilde{\mathscr{G}}(\sigma)$ is not strongly connected. In light of the discussion given prior to this theorem, the set $\{1,2,...,v\}$ can be partitioned as $\{i_1,i_2,...,i_p\}$ and $\{j_1,j_2,...,j_q\}$ such that there exists no directed edge from vertex i_α to vertex j_β in the digraph $\tilde{\mathscr{G}}(\sigma)$, for all $\alpha \in \{1,2,...,p\}$ and $\beta \in \{1,2,...,q\}$. Consider vertices $i_1,i_2,...,i_p$ in set 1 and vertices $j_1,j_2,...,j_q$ in set 2 of the bipartite graph $\mathscr{G}(\sigma)$. Denote with $\mathscr{G}_o(\sigma)$ the bipartite subgraph induced by these vertices, i.e.:

- Set 1 of $\mathscr{G}_o(\sigma)$ consists of vertices $i_1,i_2,...,i_p$ in set 1 of $\mathscr{G}(\sigma)$.
- Set 2 of $\mathscr{G}_o(\sigma)$ consists of vertices $j_1,j_2,...,j_q$ in set 2 of $\mathscr{G}(\sigma)$.
- The edges of this subgraph have been induced from the original graph $\mathscr{G}(\sigma)$.

For every $\alpha \in \{1,2,...,p\}$ and $\beta \in \{1,2,...,q\}$, since vertex i_α is not connected to vertex j_β in the digraph $\tilde{\mathscr{G}}(\sigma)$, it follows from Definitions 2 and 3 that vertex i_α in set 1 of the bipartite graph $\mathscr{G}(\sigma)$ is connected to vertex j_β in set 2 of this graph. This leads to the first observation that the induced subgraph $\mathscr{G}_o(\sigma)$ is complete bipartite. On the other hand,

$$\{1,2,...,v\} = \{i_1,i_2,...,i_p\} \cup \{j_1,j_2,...,j_q\}. \tag{3.8}$$

Now, it follows immediately from Theorem 1 that the mode σ is a DFM, due to the fact that the graph $\mathscr{G}(\sigma)$ contains a subgraph $\mathscr{G}_o(\sigma)$ with the required properties.

Proof of necessity: Assume that σ is a DFM. It is to be proved that the digraph $\tilde{\mathscr{G}}(\sigma)$ is not strongly connected. To this end, one can utilize Theorem 1 to deduce that the graph $\mathscr{G}(\sigma)$ possesses a bipartite subgraph $\mathscr{G}_o(\sigma)$ with the two properties mentioned earlier. Denote the sets of vertices of this bipartite subgraph with $\{i_1,i_2,...,i_p\}$ and $\{j_1,j_2,...,j_q\}$. Due to the properties of $\mathscr{G}_o(\sigma)$, not only is the relation (3.8) satisfied, but the following are true for every $\alpha \in \{1,2,...,p\}$ and $\beta \in \{1,2,...,q\}$:

- The first column of C_α is zero.
- The first row of B_β is zero.
- $M_{\alpha\beta}(\sigma)$ is a zero matrix.

This means that there exists no directed edge from vertex i_α to vertex j_β in the digraph $\tilde{\mathscr{G}}(\sigma)$. As a result of this observation and the relation (3.8), one can conclude that the set of vertices of $\tilde{\mathscr{G}}(\sigma)$ can be partitioned into two subsets $\{i_1,i_2,...,i_p\}$ and $\{j_1,j_2,...,j_q\}$ such that there exists no directed edge from vertices $i_1,i_2,...,i_p$ to vertices $j_1,j_2,...,j_q$. Hence, the digraph $\tilde{\mathscr{G}}(\sigma)$ is not strongly connected. ■

Theorem 2 states that checking the decentralized fixedness of σ reduces to testing the strong connectivity of the digraph $\tilde{\mathscr{G}}(\sigma)$. Fortunately, the latter problem is a very simple combinatorial problem, for which several methods have been developed. For instance, one can use Kosaraju's algorithm, which has been regarded as the simplest method for this graph problem [15]. This algorithm performs two complete traversals of the graph and the idea behind it is a depth-first search. Alternatively, Tarjan's algorithm can be employed, whose efficiency is better than Kosaraju's algorithm [16]. Another efficient algorithm, which is mostly suitable for dense graphs, is the Cheriyan/Mehlhorn/Gabow algorithm [17]. Tarjan's algorithm will be adopted in this chapter to check the strong connectivity of $\tilde{\mathscr{G}}(\sigma)$. Note that this algorithm has been implemented in the Bioinformatics Toolbox of MATLAB.

Lemma 1 *Given the matrix $M(\sigma)$, the complexity of constructing the graph $\tilde{\mathscr{G}}(\sigma)$ is at most equal to $O(n^2)$.*

Proof: Define the following:

- T is a binary $\nu \times \nu$ matrix whose (i,j) entry is equal to 0 iff the matrix $M_{ij}(\sigma)$ is equal to 0, for all $i,j \in \nu$.
- T_C is a vector of order ν whose i-th entry is equal to 0 iff the first column of C_i is equal to 0, for all $i \in \nu$.
- T_B is a vector of order ν whose j-th entry is equal to 0 iff the first row of B_j is equal to 0, for all $j \in \nu$.

It is evident that the complexity of forming the abovementioned matrices is equal to that of reading off the entries of the matrix $M(\sigma)$, the first column of C and the first row of B. This leads to the complexity order $O(n^2)+O(r)+O(m)=O(n^2)$. In the course of setting up the graph $\tilde{\mathscr{G}}(\sigma)$, the (i,j) entry of T, the i-th entry of T_C and the j-th entry of T_B are to be checked, for every $i,j \in \nu$, in order to determine if there should be an edge connecting vertex i to vertex j. The complexity of this step is at most equal to $O(\nu^2)$. The proof follows from the fact that $O(\nu^2)+O(n^2)=O(n^2)$. ■

Let the time complexity of matrix multiplication in $\mathfrak{R}^{n\times n}$ be denoted by $O(n^{\omega})$, where ω is a positive real. The following theorem presents one of the main results of this work.

Theorem 3 *Consider the decision problem "whether or not the mode σ is a DFM". This problem can be solved in $O(n^{\omega})$ time by first computing the matrix $M(\sigma)$, and then testing the strong connectivity of its associated digraph $\tilde{\mathscr{G}}(\sigma)$.*

Proof: Denote the number of edges of $\tilde{\mathscr{G}}(\sigma)$ with η. It is well-known that Tarjan's algorithm runs in $O(\nu+\eta)$ time in order to test the strong connectivity of $\tilde{\mathscr{G}}(\sigma)$ [16]. Since η is less than or equal to $\nu(\nu-1)$, the complexity of checking the connectivity of the graph $\tilde{\mathscr{G}}(\sigma)$ is at most $O(\nu^2)$. On the other hand, it is known that matrix inversion and matrix multiplication have identical time complexity exponent [18]. Since $M(\sigma)$ is computed by one matrix inversion and two matrix multiplications, $M(\sigma)$ can be found in $O(n^{\omega})$ time. Moreover, the complexity of matrix multiplication over $\mathfrak{R}^{n\times n}$ is at least $O(n^2)$, because there are n^2 entries in the matrix which must be part of any computation. Hence, the quantity ω is at least equal to 2. These results along with Lemma 1 lead to the conclusion that checking the decentralized fixedness of σ can be accomplished in $O(n^2)+O(\nu^2)+O(n^{\omega})=O(n^{\omega})$ time (note that $\nu \leq n$, by assumption). ■

Remark 1 *If the standard method of matrix multiplication is used to compute $M(\sigma)$, the time complexity of checking the decentralized fixedness of σ turns out to be $O(n^3)$, in light of Theorem 3. However, one can employ the Coppersmith-Winograd algorithm for matrix multiplication to reduce this complexity to $O(n^{2.376})$ [19].*

Corollary 1 *Consider the decision problem "whether or not the mode σ is a DFM". If there exists an algorithm for this decision problem which runs in $O(n^{\bar{\omega}})$ where $\bar{\omega} < \omega$, the algorithm must not use any of the following operations over arbitrary unstructured matrices of approximate dimension $n \times n$: matrix multiplication, matrix inversion, determinant, LUP-decomposition, computing the characteristic polynomial, orthogonal basis transformation, matrix rank.*

Proof: The proof follows from Theorem 3 and the fact that the operations pointed out in the corollary have (complexity) exponents greater than or equal to that of matrix multiplication [18]. ■

Remark 2 *In this chapter, the conventional decentralized control structure has been considered in which each local controller receives the output of one subsystem in order to generate the input signal of the same subsystem. However, there are numerous applications for which each local controller can receive the outputs of more than one subsystem (based on a given information flow structure). This control structure is usually referred to as structurally constrained control or decentralized overlapping control. The modes of the system $\mathscr{S}$ which are fixed with respect to every LTI decentralized overlapping controller are said to be decentralized overlapping fixed modes (DOFMs) [20]. It is shown in the recent paper [20] that the DOFMs of the system $\mathscr{S}$ are identical to the DFMs of a virtual system. Hence, the method proposed in this chapter can be employed to obtain the complexity of finding the DOFMs (by working on the DFMs of the virtual system).*

3.4 Numerical example

Let $\mathscr{S}$ be composed of 10 single-input single-output interconnected subsystems, with the state-space matrices given in (3.9). Notice that the matrix A is already in the canonical form (3.6) for $\sigma = 1$.

It is desired now to check whether the mode $\sigma = 1$ is a DFM of the system. To this end, the matrix $M(\sigma)$ introduced in (3.7) should be computed first. The digraph $\tilde{\mathscr{G}}(\sigma)$ constructed in terms of this matrix is depicted in Figure 3.1. This graph has 10 vertices and 47 edges. Tarjan's Algorithm can be employed to traverse all these edges and vertices in order to find the strongly connected components of this graph. This is carried out in the Bioinformatics toolbox of MATLAB using the command "graphconncomp". Vertices 1, 2 and 3 in Figure 3.1 have been colored dark blue, meaning that MATLAB has detected them as a connected component of the digraph and the remaining vertices as another component. Consequently, the

$$A=\begin{bmatrix}1&0&0&0&0&0&0&0&0&0\\0&1&1&0&1&0&1&1&1&1\\0&1&1&0&1&1&1&1&1&0\\0&1&1&1&0&0&0&0&1&0\\0&0&1&1&1&1&1&0&1&0\\0&1&1&1&0&0&0&0&1&0\\0&1&0&1&0&1&0&1&1&0\\0&1&1&0&0&1&1&1&0&0\\0&0&1&1&1&1&1&1&0&1\\0&1&1&1&1&1&0&1&0&1\end{bmatrix},\quad B=\begin{bmatrix}0&1&0&0&0&0&0&0&0&0\\1&1&-1&1&1&0&1&0&-1&1\\1&-1&-1&-1&-1&-1&-1&-1&0&1\\-1&1&0&0&-1&-1&1&0&-1&1\\1&-1&-1&0&0&1&0&-1&1&1\\0&1&-1&0&-1&-1&1&1&1&-1\\-1&0&1&-1&1&-1&0&1&0&1\\-1&-1&0&1&0&1&0&-1&0&0\\-1&1&1&-1&0&0&-1&1&1&0\\-1&1&1&-1&-1&1&1&0&-1&-1\end{bmatrix},$$

$$C=\begin{bmatrix}0&-1&1&1&-1&0&0&1&0&-1\\0&-1&1&1&1&1&0&1&0&0\\0&1&-1&0&0&-1&1&1&0&1\\0&0&0&1&0&1&-1&1&-1&1\\0&0&1&0&-1&0&-1&0&-1&0\\1&1&-1&-1&-1&1&1&0&-1&0\\0&-1&-1&-1&1&-1&1&-1&0&1\\0&0&0&-1&0&1&-1&1&-1&-1\\0&1&0&0&-1&-1&0&1&-1&0\\0&-1&-1&-1&0&0&-1&1&0&-1\end{bmatrix}, \tag{3.9}$$

$$D=\begin{bmatrix}1&-31/3&1&22&125/6&86/3&43/3&-7&-5&-107/6\\1&1&0&3&10/3&23/3&13/3&-2&-3&-10/3\\1&1&1&2&10/3&-7/3&-8/3&2&4&-4/3\\1&5/3&1&1&4/3&-7/3&1&3&1&5/3\\1&1&1&8&1&28/3&1&1&1&1\\-1/2&1&1&1&9/2&2&1&-1&2&-3/2\\-2&20/3&1&1&-41/3&1&1&4&1&1\\1&1&0&12&1&17&9&-5&1&1\\1&1&1&1&1&1&1&1&-1&-25/2\\1&1&-2&12&1&49/3&26/3&-5&1&-29/3\end{bmatrix}$$

digraph is not strongly connected, and hence $\sigma = 1$ is a DFM. Note that even though the graph $\tilde{\mathscr{G}}(\sigma)$ seems to be complex, its connectivity verification is an easy job which can be accomplished in quadratic running time (in terms of ν). In this regard, it is worth mentioning that computing the matrix $M(\sigma)$ is more involved than testing the connectivity of $\tilde{\mathscr{G}}(\sigma)$. This clearly shows the simplicity of the method proposed here.

3.5 Summary

This chapter deals with the time complexity of checking the existence of a stabilizing linear time-invariant (LTI) decentralized controller for a given LTI interconnected system. In particular, the complexity of computing the decentralized fixed modes (DFMs) of a system is studied. It is well-known that the existing deterministic methods for this problem are computationally intractable, whereas the available

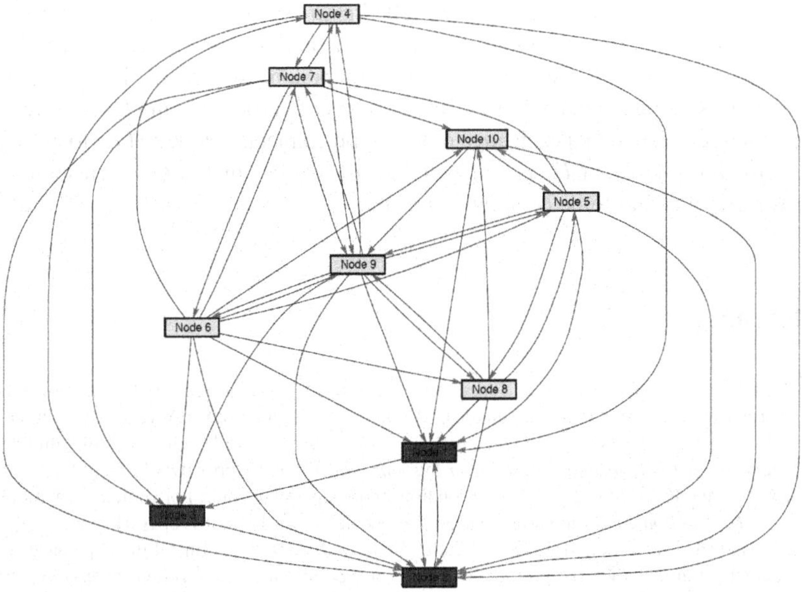

Fig. 3.1 The digraph $\tilde{\mathscr{G}}(\sigma)$ corresponding to the system used in the numerical example.

randomized numerical method is fairly simple. To investigate the true complexity of solving this problem using a deterministic algorithm, it is shown that checking whether a certain (unrepeated) mode of the system is a DFM amounts to testing the strong connectivity of some digraph. This gives rise to proving that the verification of the decentralized fixedness of a distinct mode of the system has the same time complexity as matrix multiplication and inversion.

3.6 Appendix

This part presents some classical definitions in the field of graph theory that have been used in this chapter. A *graph* G can be defined as a pair (V, E), where V is a set of vertices and E is a set of edges between the vertices such that $E \subseteq \{\{u,v\} | u, v \in V\}$. A graph $G' = (V', E')$ is a *subgraph* of G if

$$V' \subseteq V, \quad (E' \subseteq E) \wedge ((u,v) \in E' \to u, v \in V'). \tag{3.10}$$

A subgraph G' of the graph G is said to be *induced* if, for every pair of vertices u and v in G', the pair (u, v) is an edge of G' if and only if it is an edge of G. A *connected component* of the graph $\mathscr{G}$ is a subgraph in which there exists a path between every two vertices, and to which no more vertices can be added while preserving the path connectivity property. If the edges of G are ordered pairs of vertices, the graph is said to be *directed* (digraph); otherwise, it is called *undirected.* A connected component of a digraph is said to be *strongly connected* if there exists a directed path from every vertex to every other vertex within that component (subgraph). The graph G is *bipartite* if its vertices can be divided into two sets of vertices U_1 and U_2 (referred to as set 1 and set 2) such that each edge of the graph connects a vertex of U_1 to a vertex of U_2. If every vertex of U_1 is connected to all vertices of U_2, the corresponding bipartite graph is said to be *complete bipartite*.

References

1. D. M. Stipanovic, G. Inalhan, R. Teo, and C. J. Tomlin, "Decentralized overlapping control of a formation of unmanned aerial vehicles," *Automatica*, vol. 40, no. 8, pp. 1285-1296, 2004.
2. A. I. Zecevic, G. Neskovic, and D. D. Siljak, "Robust decentralized exciter control with linear feedback," *IEEE Transactions on Power Systems*, vol. 19, no. 2, pp. 1096-1103, 2004.
3. J. A. Fax and R. M. Murray, "Information flow and cooperative control of vehicle formations," *IEEE Transactions on Automatic Control*, vol. 49, no. 9, pp. 1465- 1476, 2004.
4. J. Lavaei, A. Momeni, and A. G. Aghdam, "Spacecraft formation control in deep space with reduced communication requirement", *IEEE Transactions on Control System Technology*, vol. 16, no. 2, pp. 268-278, 2008.
5. K. L. Kosmatopoulos, E. B. Ioannou, and P. A. Ryaciotaki-Boussalis, "Large segmented telescopes: centralized decentralized and overlapping control designs," *IEEE Control Systems Magazine*, vol. 20, no. 5, pp. 59-72, 2000.
6. D. D. Siljak, Decentralized control of complex systems, Cambridge: Academic Press, 1991.
7. S. H. Wang and E. J. Davison, "On the stabilization of decentralized control systems," *IEEE Transactions on Automatic Control*, vol. 18, no. 5, pp. 473-478, 1973.
8. B. L. O. Anderson and D. J. Clements, "Algebraic characterizations of fixed modes in decentralized systems," *Automatica*, vol. 17, no. 5, pp. 703-712, 1981.
9. B. L. O. Anderson, "Transfer function matrix description of decentralized fixed modes," *IEEE Transactions on Automatic Control*, vol. 27, no. 6, pp. 1176-1182, 1982.
10. E. J. Davison and T. N. Chang, "Decentralized stabilization and pole assignment for general proper systems," *IEEE Transactions on Automatic Control*, vol. 35, no. 6, pp. 652-664, 1990.
11. J. Lavaei and A. G. Aghdam, "A graph theoretic method to find decentralized fixed modes of LTI systems," *Automatica*, vol. 43, no. 12, pp. 2129-2133, 2007.
12. E. J. Davison and S. H. Wang, "A characterization of decentralized fixed modes in terms of transmission zeros," *IEEE Transactions on Automatic Control*, vol. 30, no. 1, pp. 81-82, 1985.
13. E. J. Davison, "Decentralized stabilization and regulation in large multivariable systems," in *Directions in Large Scale Systems*, Y. C. Ho and S. K. Mitter, Eds. New York: Plenum, pp. 303-323, 1976.
14. A. G. Aghdam and E. J. Davison, "Discrete-time control of continuous systems with approximate decentralized fixed modes," *Automatica*, vol. 44, no. 1, pp. 75-87, 2008.
15. T. H. Cormen, C. E. Leiserson, R. L. Rivest and C. Stein, "Introduction to Algorithms," The MIT Press, 2nd edition, 2001.
16. R. Tarjan, "Depth-first search and linear graph algorithms," *SIAM Journal on Computing*, vol. 1, no. 2, pp. 146-160, 1972.
17. J. Cheriyan and K. Mehlhorn, "Algorithms for dense graphs and networks on the random access computer," *Algorithmica*, vol. 15, no. 6, pp. 521-549, 1996.
18. P. Bürgisser, M. Clausen and M. A. Shokrollahi, "Algebraic complexity theory," Springer, 1997.
19. D. Coppersmith and S. Winograd, "Matrix multiplication via arithmetic progressions," *Journal of Symbolic Computation*, vol. 9, no. 3, pp. 251-280, 1990.

20. J. Lavaei and A. G. Aghdam, "Control of continuous-Time LTI systems by means of structurally constrained controllers," *Automatica*, vol. 44, no. 1, pp. 141-148, 2008.

Chapter 4
Decentralized Overlapping Control: Stabilizability and Pole-Placement

4.1 Introduction

In the past three decades, the problem of decentralized control has been thoroughly investigated in the literature, and a variety of its aspects are studied [1, 2, 3]. More recently, the problem of decentralized overlapping control has attracted several researchers [4, 5]. The decentralized overlapping control is fundamentally used in two cases:

i) when the subsystems of a system (referred to as overlapping subsystems) share some states [6, 7, 8]. In this case, it is usually desired that the structure of the controller matches the overlapping structure of the system [8];
ii) when there are some limitations on the availability of the states. In this case, only certain outputs of the system are available for constructing each control signal.

The control constraint in both cases discussed above can be represented by a binary information flow matrix. For instance, when this matrix is block diagonal with the entries of the main diagonal blocks all equal to 1, the control structure is decentralized, and when all of its entries are 1, the controller is centralized. One particular structural constraint for the controller, which is investigated intensively in the literature, corresponds to an information flow matrix whose entries on the main diagonal blocks, as well as the last block column and the last block row are all equal to 1. This is often referred to as bordered block-diagonal structure (BBD) or block array structure (BAS), and has found several practical applications [8, 9, 10]. In general, for an interconnected system with a given information flow matrix, the following open questions are of main interest in the literature:

1. Does there exist a stabilizing static output feedback controller for the system?
2. Does there exist a linear time-invariant (LTI) controller to stabilize the system, if there is no static one?
3. How can a static or dynamic LTI controller be found such that a predefined quadratic performance index is minimized?

4. Can the poles of the system be placed at any arbitrary locations, when there exists a LTI stabilizing controller for the system?
5. Can the system be stabilized by a non-LTI controller when a LTI stabilizing controller does not exist?

The first three questions have been addressed in the literature in the decentralized overlapping control framework. This is accomplished by using a transformation which expands the structure of the system such that the resultant control configuration is decentralized. Then, by using the existing design techniques, the desired decentralized controller is obtained for the expanded system. The last step of the design is to contract the controller obtained in order to make it suitable for the original system. This approach is substantially useful, when the structure of the system itself is overlapping as well because in that case, the subsystems of the expanded system are disjoint [11]. Nevertheless, one of the shortcomings of this method is that the expanded system is inherently uncontrollable, and thus, this design approach may not be useful in general. This problem has been addressed in several papers, e.g. see [8], [4]. Furthermore, the contraction of the designed controller can cause some problems in general. Although a large number of conditions for contraction are presented, finding a proper contraction is still an open problem [8]. In addition, it is often assumed that a static state feedback controller (as opposed to a general output feedback controller) is to be designed, which may not be suitable in practical applications. In the special case of a BAS control design, a number of methods have been proposed in the literature, including an optimal control design technique [9, 10]. The other existing methods for designing a BAS or an overlapping (or structurally constrained) controller often present some sufficient conditions in the form of LMI, and fail to address some of the important questions discussed above [8, 12, 13]. Furthermore, these methods assume that the system is strictly proper, while the generalization of the methods to general proper systems is not straightforward. The present work addresses the problem of designing a structurally constrained controller, and is aimed to answer the open questions discussed above, for any LTI system with any arbitrary information flow structure.

4.2 Problem formulation

Consider a LTI interconnected system $\mathscr{S}$ consisting of ν subsystems with the following state-space representation:

$$\begin{aligned}\dot{x}(t) &= Ax(t) + \sum_{i=1}^{\nu} B_i u_i(t)\\ y_i(t) &= C_i x(t) + \sum_{j=1}^{\nu} D_{ij} u_j(t), \quad i \in \bar{\nu} := \{1,2,...,\nu\}\end{aligned} \tag{4.1}$$

where $x(t) \in \Re^n$ is the state, and $u_i(t) \in \Re^{m_i}$ and $y_i(t) \in \Re^{r_i}$, $i \in \bar{\nu}$, are the input and the output of the ith subsystem S_i, respectively. Define the following matrices:

$$B := \begin{bmatrix} B_1 \; B_2 \; \cdots \; B_\nu \end{bmatrix}, \quad C := \begin{bmatrix} C_1 \\ C_2 \\ \vdots \\ C_\nu \end{bmatrix}, \quad D := \begin{bmatrix} D_{11} & \cdots & D_{1\nu} \\ \vdots & \ddots & \vdots \\ D_{\nu 1} & \cdots & D_{\nu\nu} \end{bmatrix} \tag{4.2}$$

Define also:

$$m := \sum_{i=1}^{\nu} m_i, \quad r := \sum_{i=1}^{\nu} r_i \tag{4.3}$$

It is desired to stabilize the system $\mathscr{S}$ by using a structurally constrained controller. These constraints determine which outputs y_j $(j \in \bar{\nu})$ are available to construct any specific input u_i $(i \in \bar{\nu})$ of the system. In order to simplify the formulation of the control constraint, a matrix $\mathscr{K}$ with binary elements is defined, where its (i,j) block entry, $i,j \in \bar{\nu}$, is a $m_i \times r_j$ matrix whose elements are all equal to 1 if the output y_j can contribute to the construction of the input u_i, and is a $m_i \times r_j$ zero matrix otherwise. The matrix $\mathscr{K}$ represents the control constraint, and will be referred to as the information flow matrix.

To represent the structural constraint of the system, the corresponding information flow matrix is enclosed in parentheses throughout the chapter, if necessary. For instance, $\mathscr{S}(\mathscr{K})$ indicates that the structure of the controller to be designed for the system $\mathscr{S}$ is to comply with the information flow matrix $\mathscr{K}$.

In the special case, when the entries of the matrix $\mathscr{K}$ are all equal to 1, the corresponding controller is centralized, and when $\mathscr{K}$ is block diagonal, the corresponding controller is decentralized. Throughout this chapter, the term "decentralized controller" is referred to the set of local controllers for an interconnected system with a block diagonal information flow matrix.

It is to be noted that in the case of a block diagonal matrix $\mathscr{K}$, one can use the existing methods, e.g. [2], to find the decentralized fixed modes (DFM) of the system, if any. Then, if the system does not have any DFM in the closed right-half plane (RHP), one can use a LTI decentralized controller to stabilize it and place those modes which are not fixed, in any arbitrary location in the complex plane. Furthermore, the system can still be stabilized in the presence of unstable DFMs, as long as they are not quotient fixed modes (QFM) [3]. A system with unstable QFMs cannot be stabilized by using any type of controller, i.e., nonlinear and time-varying. However, there is no necessary and sufficient condition for the existence of a general stabilizing controller, when $\mathscr{K}$ is not block diagonal. This problem will be addressed in the following sections. It is to be noted that to avoid trivial cases (i.e., standard decentralized and centralized systems), the matrix $\mathscr{K}$ will hereafter be assumed not to be block diagonal, and to have at least one zero block.

4.3 Computing the transformation matrices

Definition 1 *Consider two arbitrary systems $\mathscr{S}_{d_1}$ and $\mathscr{S}_{d_2}$ associated with the information flow matrices $\mathscr{K}_{d_1}$ and $\mathscr{K}_{d_2}$, where $\mathscr{S}_{d_1}$ and $\mathscr{S}_{d_2}$ are of the same order and have the same initial state. Let $\mathbf{M}$ denote a given set of controllers. The systems $\mathscr{S}_{d_1}(\mathscr{K}_{d_1})$ and $\mathscr{S}_{d_2}(\mathscr{K}_{d_2})$ are called* analogous *with respect to $\mathbf{M}$ if for any controller K_{d_1} in $\mathbf{M}$ complying with the information flow matrix $\mathscr{K}_{d_1}$, there also exists a controller K_{d_2} in $\mathbf{M}$ complying with the information flow matrix $\mathscr{K}_{d_2}$ (and vice versa), such that the state of the system $\mathscr{S}_{d_1}$ under the controller K_{d_1} is equivalent to the state of $\mathscr{S}_{d_2}$ under K_{d_2}.*

The motivation for introducing the notion of analogous systems is that given a system $\mathscr{S}(\mathscr{K})$ with any general information flow structure $\mathscr{K}$, it is desired to find an analogous system with a decentralized (i.e. block diagonal) information flow structure. It is to be noted there are several efficient methods for design of decentralized controllers. Thus, the problem reduces to designing a proper decentralized controller, and finding a transformation to change the block-diagonal structure of the controller to the desired structure for the original system $\mathscr{S}(\mathscr{K})$. This is an indirect method of design, which unlike the existing indirect approaches aims to identify the fixed modes with respect to a structurally constrained controller. This section presents some transformation matrices which will later be used to construct systems *analogous* to the system $\mathscr{S}(\mathscr{K})$.

Define the control *interaction* structure $\mathbf{K}$ as a matrix whose (i,j) block entry, $i,j \in \bar{\mathrm{v}}$, is a $m_i \times r_j$ matrix denote by k_{ij} if the output of the jth subsystem can contribute to the construction of the input of the ith subsystem, and is a $m_i \times r_j$ zero matrix otherwise. Note that k_{ij} represents a component of the controller, which transforms the output of the jth subsystem to the input of the ith subsystem. Note also that the interaction structure matrix $\mathbf{K}$ not only conveys the information of the matrix $\mathscr{K}$, but also labels the control components.

Procedure 1 *Construct the graph $\mathscr{G}$ as follows:*

1. *Define two sets of v vertices. Label the sets as set 1 and set 2, and the vertices in each set as vertex 1 to vertex v.*
2. *For any $i,j \in \bar{\mathrm{v}}$, connect the ith vertex of set 1 to the jth vertex of set 2 with an edge, if the (i,j) block entry of $\mathscr{K}$ is not a zero matrix, i.e., if the output of the jth subsystem can contribute to the construction of the input of the ith subsystem. Label this edge with k_{ij}.*

As an example, consider a system consisting of four subsystems with the following control interaction structure matrix:

$$\mathbf{K} = \begin{bmatrix} k_{11} & 0 & 0 & 0 \\ k_{21} & k_{22} & 0 & k_{24} \\ k_{31} & 0 & k_{33} & 0 \\ 0 & k_{42} & 0 & k_{44} \end{bmatrix} \tag{4.4}$$

The graph $\mathscr{G}$ corresponding to the matrix $\mathbf{K}$ given above is depicted in Figure 4.1.

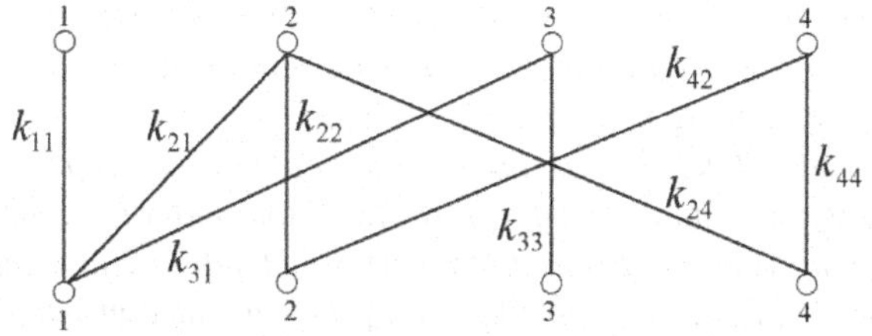

Fig. 4.1 The graph $\mathscr{G}$ corresponding to the matrix $\mathbf{K}$ given by (4.4).

Procedure 2 *Partition the graph $\mathscr{G}$ into a set of complete bipartite subgraphs such that each edge of the graph $\mathscr{G}$ appears in only one of the subgraphs. It is to be noted that this partition may require some of the vertices of the graph $\mathscr{G}$ to appear in several subgraphs.*

It can be easily verified that Procedure 2 does not necessarily lead to a unique graph. Denote all the graphs which can be obtained through this procedure, with $\mathscr{G}_1, \mathscr{G}_2, ..., \mathscr{G}_l$. Without loss of generality, assume that $\mathscr{G}_1$ and $\mathscr{G}_l$ are the ones with the following properties:

- $\mathscr{G}_1$ is obtained by considering any vertex in set 1 of the graph $\mathscr{G}$ along with all of the vertices in set 2 connected to that vertex as a complete bipartite graph.
- $\mathscr{G}_l$ is obtained by considering any edge in the graph $\mathscr{G}$ as a complete bipartite graph.

As an example, consider again the graph $\mathscr{G}$ sketched in Figure 4.1. The graph $\mathscr{G}_2$ for this graph can be considered as the one depicted in Figure 4.2 (note that this graph is denoted by $\mathscr{G}_2$ instead of $\mathscr{G}_1$, because it does not satisfy the property of $\mathscr{G}_1$ described above). It is obvious from Figure 4.2 that, in this particular example, vertices 2 and 3 of the first set of vertices of $\mathscr{G}$ are repeated twice in $\mathscr{G}_2$.

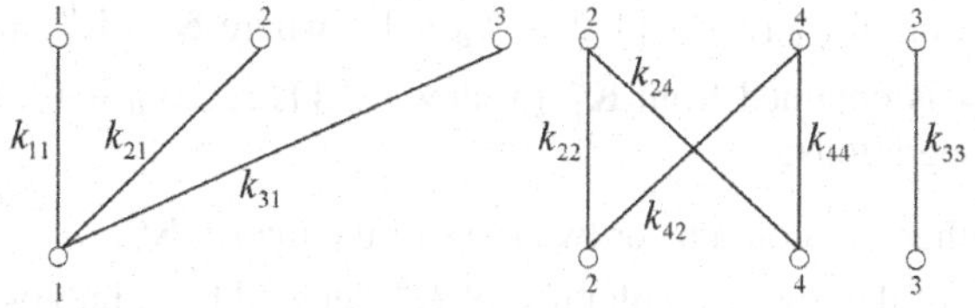

Fig. 4.2 A decentralized graph $\mathscr{G}_2$ obtained from the graph $\mathscr{G}$ in Figure 4.1.

The following procedure can be used to construct the matrix $\mathbf{K}_\mu$ corresponding to the graph $\mathscr{G}_\mu$ for any $\mu \in \bar{l} := \{1,2,...,l\}$.

Procedure 3 *Label the complete bipartite subgraphs of $\mathscr{G}_\mu$ ($\mu \in \bar{l}$) as subgraphs 1 to ν_μ. Consider subgraph number σ ($\forall\ \sigma \in \{1,2,...,\nu_\mu\}$). Label those vertices of* *this subgraph which belong to set 1 as vertex* $1,...,\eta_\sigma^\mu$. *This group of vertices will be referred to as subset 1 (corresponding to subgraph number* σ*). Similarly, label those vertices which belong to set 2 of this subgraph as vertex* $1,...,\bar{\eta}_\sigma^\mu$, *and define subset 2 accordingly. Define* $\mathbf{K}_\mu$ *as a block diagonal matrix, where its* (σ,σ) *block entry,* $\sigma = 1,...,\nu_\mu$, *is a matrix itself, whose* (i,j) *block entry is equal to the gain of the edge connecting vertex i of subset 1 to vertex j of subset 2 in subgraph number* σ *of* $\mathscr{G}_\mu$, *for any* $i \in \{1,...,\eta_\sigma^\mu\}$, $j \in \{1,...,\bar{\eta}_\sigma^\mu\}$. *Denote the dimension of the* (σ,σ) *block entry of* $\mathbf{K}_\mu$ *with* $m_\sigma^\mu \times r_\sigma^\mu$, *for* $\sigma = 1,2,...,\nu_\mu$, *and the dimension of* $\mathbf{K}_\mu$ *with* $m^\mu \times r^\mu$.

Using Procedure 3 and for a particular numbering of vertices in each subgraph of $\mathscr{G}_2$ in Figure 4.2, the following block diagonal matrix $\mathbf{K}_2$ is obtained:

$$\mathbf{K}_2 = \begin{pmatrix} k_{11} & 0 & 0 & 0 \\ k_{21} & 0 & 0 & 0 \\ k_{31} & 0 & 0 & 0 \\ 0 & k_{22} & k_{24} & 0 \\ 0 & k_{42} & k_{44} & 0 \\ 0 & 0 & 0 & k_{33} \end{pmatrix} \tag{4.5}$$

Remark 1 *It can be easily concluded from Procedures 1, 2 and 3, that there exists an onto mapping between the nonzero block entries of the matrix* $\mathbf{K}_\mu$ *and those of the matrix* $\mathbf{K}$ *for any* $\mu \in \bar{l}$.

Theorem 1 *There exist constant matrices* Φ_μ *and* $\bar{\Phi}_\mu$ *satisfying the following relation:*

$$\mathbf{K} = \Phi_\mu \mathbf{K}_\mu \bar{\Phi}_\mu \tag{4.6}$$

for any $\mu \in \bar{l}$.

Proof. It is straightforward to show (by using Procedures 1, 2 and 3) that the matrix $\mathbf{K}_\mu$ can alternatively be constructed from $\mathbf{K}$ through a sequence of $L_\mu - 1$ operations (where L_μ is a finite number), such that the matrix $\mathbf{K}_{j+1}^\mu$ is formed in terms of $\mathbf{K}_j^\mu$ in the jth operation, for any $j \in \{1,2,...,L_\mu - 1\}$, where $\mathbf{K} = \mathbf{K}_1^\mu$ and $\mathbf{K}_\mu = \mathbf{K}_{L_\mu}^\mu$. Moreover, $\mathbf{K}_{j+1}^\mu$ is obtained from $\mathbf{K}_j^\mu$ for any $j \in \{1,2,...,L_\mu - 1\}$, by one of the following two operations:

1. Swapping either two columns or two rows of the matrix $\mathbf{K}_j^\mu$.
2. Splitting one of the rows (or columns) of $\mathbf{K}_j^\mu$ denoted by v, into two row vectors v_1 and v_2, i.e., $v = [v_1 \;\; v_2]$ (or $v = [v_1 \;\; v_2]'$). Then, replacing that row (or column) with $[v_1 \;\; 0]$ (or $[v_1 \;\; 0]'$), where 0 represents a zero row vector, and inserting another row (or column) equal to $v = [0 \;\; v_2]$ (or $v = [0 \;\; v_2]'$) into the matrix.

It is desired now to prove for any $j \in \{1,...,L_\mu - 1\}$, that there exist matrices Φ_j^μ and $\bar{\Phi}_j^\mu$ such that $\mathbf{K}_j^\mu = \Phi_j^\mu \mathbf{K}_{j+1}^\mu \bar{\Phi}_j^\mu$.

1. Assume that $\mathbf{K}_{j+1}^{\mu}$ is derived from $\mathbf{K}_j^{\mu}$ by swapping its gth and q^{th} columns. It is straightforward to show in this case, that the matrices Φ_j^{μ} and $\bar{\Phi}_j^{\mu}$ will be as follows:

   a. Φ_j^{μ} is an identity matrix, whose dimension is equal to the number of rows of $\mathbf{K}_j^{\mu}$.

   b. $\bar{\Phi}_j^{\mu}$ is derived from an identity matrix, whose dimension is equal to the number of columns of $\mathbf{K}_j^{\mu}$, by setting the (g,g) and (q,q) entries of this identity matrix to zero, and setting its (g,q) and (q,g) entries to one.

   It is to be noted that if two rows of $\mathbf{K}_j^{\mu}$ instead of two columns are swapped, then the procedures to obtain the matrices Φ_j^{μ} and $\bar{\Phi}_j^{\mu}$ should also be swapped.

2. Assume that one of the columns of the matrix $\mathbf{K}_j^{\mu}$ is split into two columns as described before (note that the case of row split can be carried out in a similar manner). For instance, suppose that $\mathbf{K}_j^{\mu}$ is as follows:

$$\mathbf{K}_j^{\mu} = \begin{bmatrix} M_1 \; m_2 \; M_3 \\ M_4 \; m_5 \; M_6 \end{bmatrix} \tag{4.7}$$

   where:

$$\begin{aligned} M_1 \in \Re^{g_1 \times q_1}, \;\; m_2 \in \Re^{g_1 \times 1}, \;\; M_3 \in \Re^{g_1 \times q_2}, \\ M_4 \in \Re^{g_2 \times q_1}, \;\; m_5 \in \Re^{g_2 \times 1}, \;\; M_6 \in \Re^{g_2 \times q_2}, \end{aligned} \tag{4.8}$$

   In addition, consider:

$$\mathbf{K}_{j+1}^{\mu} = \begin{bmatrix} M_1 & m_2 & 0_{g_1 \times 1} & M_3 \\ M_4 & 0_{g_2 \times 1} & m_5 & M_6 \end{bmatrix} \tag{4.9}$$

   It can be easily verified that $\Phi_j^{\mu} = I_{g_1+g_2}$, and:

$$\bar{\Phi}_j^{\mu} = \begin{bmatrix} I_{q_1} & 0_{q_1 \times 1} & 0_{q_1 \times q_2} \\ 0_{1 \times q_1} & 1 & 0_{1 \times q_2} \\ 0_{1 \times q_1} & 1 & 0_{1 \times q_2} \\ 0_{q_2 \times q_1} & 0_{q_2 \times 1} & I_{q_2} \end{bmatrix} \tag{4.10}$$

   Note that $0_{q \times g}$ and I_g represent the $q \times g$ zero matrix and the $g \times g$ identity matrix, respectively, for any $g, q \geq 1$.

Hence, it is shown that for each of the two operations discussed earlier, there exist the matrices Φ_j^{μ} and $\bar{\Phi}_j^{\mu}$, which satisfy the aforementioned property. The matrices Φ_μ and $\bar{\Phi}_\mu$ can now be obtained from the following equations:

$$\Phi_\mu = \Phi_1^{\mu} \Phi_2^{\mu} \cdots \Phi_{(L_\mu - 1)}^{\mu}, \quad \bar{\Phi}_\mu = \bar{\Phi}_{(L_\mu - 1)}^{\mu} \bar{\Phi}_{(L_\mu - 2)}^{\mu} \cdots \bar{\Phi}_1^{\mu} \tag{4.11}$$

■

Theorem 1 states that there exist matrices Φ_μ and $\bar{\Phi}_\mu$ for the matrix $\mathbf{K}_\mu$ ($\mu \in \bar{l}$) derived from $\mathbf{K}$ using Procedures 1, 2 and 3, such that they satisfy the equation (4.6). However, since the proof of Theorem 1 relies on a sequence of matrices, the proposed procedure may not be efficient to compute Φ_μ and $\bar{\Phi}_\mu$ for an information flow matrix with a large number of block entries. The following theorem presents a more efficient approach to obtain Φ_μ and $\bar{\Phi}_\mu$.

Theorem 2 *Choose at least one nonzero block entry from each block column and each block row of* $\mathbf{K}_\mu$, $\mu \in \bar{l}$, *and let them be denoted by* $k_{i_1 j_1}, k_{i_2 j_2}, \ldots, k_{i_p j_p}$. *Suppose that* $k_{i_q j_q}$, $q = 1,2,\ldots,p$, *is the* (i'_q, j'_q) *block entry of the matrix* $\mathbf{K}_\mu$. *Denote the* h_1^{th} *block column of* Φ_μ *and the* h_2^{th} *block row of* $\bar{\Phi}_\mu$ *with* Π_{h_1} *and* $\bar{\Pi}_{h_2}$, *respectively, for* $h_1 = 1,2,\ldots,m^\mu$, $h_2 = 1,2,\ldots,r^\mu$ *(note that* m^μ *and* r^μ *are defined in procedure 3). Then:*

$$\Pi_{i'_q} = \begin{bmatrix} 0_{m_1 \times m_{i_q}} \\ 0_{m_2 \times m_{i_q}} \\ \vdots \\ 0_{m_{(i_q-1)} \times m_{i_q}} \\ I_{m_{i_q}} \\ 0_{m_{(i_q+1)} \times m_{i_q}} \\ \vdots \\ 0_{m_v \times m_{i_q}} \end{bmatrix}, \quad \bar{\Pi}_{j'_q} = \begin{bmatrix} 0_{r_{j_q} \times r_1} \\ 0_{r_{j_q} \times r_2} \\ \vdots \\ 0_{r_{j_q} \times r_{(j_q-1)}} \\ I_{r_{j_q}} \\ 0_{r_{j_q} \times r_{(j_q+1)}} \\ \vdots \\ 0_{r_{j_q} \times r_v} \end{bmatrix}^T \tag{4.12}$$

for any $q \in \{1,2,\ldots,p\}$.

Proof. It is shown in Theorem 1 that the matrices Φ_μ and $\bar{\Phi}_\mu$ exist to satisfy the equation (4.6). As a result, this equation holds for any arbitrary values for the block entries $k_{\sigma_1\sigma_2}$, $\sigma_1, \sigma_2 \in \bar{v}$. Replace all block entries $k_{\sigma_1\sigma_2}$'s in the equation (4.6), except $k_{i_q j_q}$, with zero matrices. It can be concluded from (4.6) that:

$$\tilde{\mathbf{K}}_{i_q j_q} = \Pi_{i'_q} k_{i_q j_q} \bar{\Pi}_{j'_q} \tag{4.13}$$

where $\tilde{\mathbf{K}}_{i_q j_q}$ is obtained from $\mathbf{K}$ by replacing all of its block entries with zero matrices, except for its (i_q, j_q) block entry $k_{i_q j_q}$. The proof follows immediately from the equation (4.13). ■

It is to be noted that the matrices Φ_μ and $\bar{\Phi}_\mu$ are uniquely determined. To illustrate the method proposed in Theorem 2, consider again the matrix $\mathbf{K}_2$ given by (4.5), which is obtained from $\mathbf{K}$ in (4.4), and assume that the subsystems of the original system are all single-input single-output (SISO). As the first step in computing Φ_2 and $\bar{\Phi}_2$, choose some of the nonzero entries of $\mathbf{K}_2$, such that at least one entry from each column and each row of $\mathbf{K}$ is included. Let these entries be $k_{11}, k_{21}, k_{31}, k_{22}, k_{44}$, and k_{33}. The position of these entries in the matrix $\mathbf{K}_2$ are $(1,1), (2,1), (3,1), (4,2), (5,3), (6,4)$, respectively. Using Theorem 2, one can obtain the matrices Φ_2 and $\bar{\Phi}_2$ for this example, as follows:

$$\Phi_2 = \begin{bmatrix} 1\,0\,0\,0\,0\,0 \\ 0\,1\,0\,1\,0\,0 \\ 0\,0\,1\,0\,0\,1 \\ 0\,0\,0\,0\,1\,0 \end{bmatrix}, \quad \bar{\Phi}_2 = \begin{bmatrix} 1\,0\,0\,0 \\ 0\,1\,0\,0 \\ 0\,0\,0\,1 \\ 0\,0\,1\,0 \end{bmatrix} \tag{4.14}$$

It is very easy to verify that these matrices satisfy the relation (4.6).

4.4 Linear time-invariant control law

In this section, it is desired to find conditions for the existence of a stabilizing LTI controller for the system $\mathscr{S}(\mathscr{K})$. Furthermore, a procedure is given to achieve pole placement using a LTI control law. Design of a structurally constrained linear-quadratic optimal controller is then studied.

4.4.1 Pole placement

Definition 2 *Define $\mathscr{S}_\mu$, $\mu \in \bar{I}$, as an interconnected system with the following state-space representation:*

$$\begin{aligned} \dot{\mathbf{x}}_\mu(t) &= A\mathbf{x}_\mu(t) + \mathbf{B}^\mu \mathbf{u}_\mu(t) \\ \mathbf{y}_\mu(t) &= \mathbf{C}^\mu \mathbf{x}_\mu(t) + \mathbf{D}^\mu \mathbf{u}_\mu(t) \end{aligned} \tag{4.15}$$

where the system parameters are related to the state-space matrices of the system $\mathscr{S}$ given by (4.1), as shown below:

$$\mathbf{B}^\mu = B\Phi_\mu, \quad \mathbf{C}^\mu = \bar{\Phi}_\mu C, \quad \mathbf{D}^\mu = \bar{\Phi}_\mu D \Phi_\mu \tag{4.16}$$

$\mathbf{u}_\mu(t) \in \Re^{m^\mu}$ *and* $\mathbf{y}_\mu(t) \in \Re^{r^\mu}$ *are the input and the output of $\mathscr{S}_\mu$, respectively, and* $\mathbf{x}_\mu(0) = x(0)$. *For any $\mu \in \bar{I}$, define the information flow matrix $\mathscr{K}_\mu$ for the system $\mathscr{S}_\mu$ as a matrix obtained from $\mathbf{K}_\mu$ by replacing its nonzero block entry k_{ij}, with a $m_i \times r_j$ matrix whose entries are all equal to one, for any $i, j \in \bar{\nu}$.*

Theorem 3 *For any $\mu \in \bar{I}$, the systems $\mathscr{S}_\mu(\mathscr{K}_\mu)$ and $\mathscr{S}(\mathscr{K})$ are* analogous *with respect to the set of all LTI controllers.*

Proof. Denote the transfer function matrix of any nonzero control component k_{ij} with $K_{ij}(s)$, $i, j \in \bar{\nu}$ (the dimension of $K_{ij}(s)$ is the same as k_{ij} but the function itself is yet to be designed). Replace the block k_{ij} with $K_{ij}(s)$ in the matrices $\mathbf{K}$ and $\mathbf{K}_\mu$ for any $i, j \in \bar{\nu}$, and denote the resultant control transfer function matrices with $K(s)$ and $K_\mu(s)$, respectively. It can be easily concluded from Theorem 1 that:

$$K(s) = \Phi_\mu K_\mu(s) \bar{\Phi}_\mu \tag{4.17}$$

Assume the control transfer function matrix $K(s)$ is such that the matrix $I_r - DK(s)$ is nonsingular. It is known that the state of the system $\mathscr{S}$ under the controller $K(s)$ satisfies the following equation:

$$X(s) = \left(sI_n - A - BK(s)\,(I_r - DK(s))^{-1}C\right)^{-1} x(0) \tag{4.18}$$

On the other hand, it can be easily verified that $I_{r^\mu} - \bar{\Phi}_\mu D\Phi_\mu K_\mu(s)$ is nonsingular due to the assumption $\det(I_r - DK(s)) \neq 0$. Similarly, the state of the system $\mathscr{S}_\mu$ under the controller $K_\mu(s)$ can be obtained as follows:

$$\mathbf{X}_\mu(s) = \left(sI_n - A - \mathbf{B}^\mu K_\mu(s)\left(I_{r^\mu} - \mathbf{D}^\mu K_\mu(s)\right)^{-1}\mathbf{C}^\mu\right)^{-1} \mathbf{x}_\mu(0) \tag{4.19}$$

Furthermore, using the equations (4.16) and (4.17), one can write:

$$\begin{aligned} BK(s)(I_r - DK(s))^{-1}C &= B\Phi_\mu K_\mu(s)\bar{\Phi}_\mu\left(I_r - D\Phi_\mu K_\mu(s)\bar{\Phi}_\mu\right)^{-1}C \\ &= B\Phi_\mu K_\mu(s)\left(I_{r^\mu} - \bar{\Phi}_\mu D\Phi_\mu K_\mu(s)\right)^{-1}\bar{\Phi}_\mu C \\ &= \mathbf{B}^\mu K_\mu(s)\left(I_{r^\mu} - \mathbf{D}^\mu K_\mu(s)\right)^{-1}\mathbf{C}^\mu \end{aligned} \tag{4.20}$$

The proof follows from the relations (4.18), (4.19), and (4.20). ■

Corollary 1 *For any $\mu \in \bar{l}$, the systems $\mathscr{S}_\mu(\mathscr{K}_\mu)$ and $\mathscr{S}(\mathscr{K})$ are* analogous *with respect to the set of all continuous-time* static *controllers.*

Proof. The proof is omitted due to its similarity to the proof of Theorem 3. ■

Remark 2 *It can be easily concluded from Theorem 3 and Corollary 1 that all of the systems $\mathscr{S}(\mathscr{K})$, $\mathscr{S}_1(\mathscr{K}_1), \mathscr{S}_2(\mathscr{K}_2), \ldots, \mathscr{S}_l(\mathscr{K}_l)$ are* analogous *with respect to the set of continuous-time dynamic LTI controllers, as well as the set of continuous-time static controllers. As a result, in order to design a continuous-time dynamic (or static) LTI controller for the system $\mathscr{S}$ with respect to the information flow structure $\mathscr{K}$ to achieve any design objective (such as pole placement), one can equivalently design a continuous-time LTI controller for the system $\mathscr{S}_\mu$, $\mu \in \bar{l}$, with respect to the information flow structure $\mathscr{K}_\mu$, to attain the same objective. The mapping between the components of* **K** *and* $\mathbf{K}_\mu$ *(derived from the equation (4.6)) can then be used to find the corresponding controller for the system $\mathscr{S}(\mathscr{K})$. The important advantage of this indirect design procedure is that the information flow structure $\mathscr{K}_\mu$ is block diagonal, and hence the problem is converted to the conventional decentralized control design problem, which can be handled by the existing methods [2, 14].*

The question arises now as which of the systems $\mathscr{S}_1, \mathscr{S}_2, \ldots, \mathscr{S}_l$ is more appropriate to be employed for the aforementioned control design procedure. It is to be noted that all of these systems are *analogous*, and hence possess similar characteristics in terms of output performance. However, a smart choice of system here is of crucial importance in terms of simplifying the control design problem. This will be discussed in detail later.

Partition now the matrices $\mathbf{B}^\mu, \mathbf{C}^\mu$ and $\mathbf{D}^\mu$, $\mu \in \bar{l}$, as follows:

$$\mathbf{B}^\mu = \left[\mathbf{B}_1^\mu \; \mathbf{B}_2^\mu \; \cdots \; \mathbf{B}_{\nu_\mu}^\mu\right], \quad \mathbf{C}^\mu = \begin{bmatrix} \mathbf{C}_1^\mu \\ \mathbf{C}_2^\mu \\ \cdots \\ \mathbf{C}_{\nu_\mu}^\mu \end{bmatrix}, \quad \mathbf{D}^\mu = \begin{bmatrix} \mathbf{D}_{1,1}^\mu & \cdots & \mathbf{D}_{1,\nu_\mu}^\mu \\ \vdots & \ddots & \vdots \\ \mathbf{D}_{\nu_\mu,1}^\mu & \cdots & \mathbf{D}_{\nu_\mu,\nu_\mu}^\mu \end{bmatrix} \tag{4.21}$$

where:

$$\mathbf{B}_i^\mu \in \Re^{m_i^\mu}, \quad \mathbf{C}_i^\mu \in \Re^{r_i^\mu}, \quad \mathbf{D}_{ij}^\mu \in \Re^{r_i^\mu \times m_j^\mu} \tag{4.22}$$

for any $i, j \in \{1, 2, ..., \nu_\mu\}$. It is to be noted that m_i^μ and r_i^μ are defined in Procedure 3.

Theorem 4 *Consider an arbitrary region $\mathscr{R}$ in the complex plane. There exists a LTI controller for the system $\mathscr{S}(\mathscr{K})$ to place all modes of the resultant closed-loop system inside the region $\mathscr{R}$, except for those modes which are DFMs of the system $\mathscr{S}_\mu$ with respect to $\mathscr{K}_\mu$, $\mu \in \bar{l}$.*

Proof. As pointed out in Remark 2, the systems $\mathscr{S}(\mathscr{K})$ and $\mathscr{S}_\mu(\mathscr{K}_\mu)$ are equivalent in terms of pole placement capabilities. On the other hand, it results from the definition of DFM [1] that all of the modes of the system $\mathscr{S}_\mu(\mathscr{K}_\mu)$ except for its DFMs can be placed arbitrarily by using a proper LTI controller. This completes the proof. ■

Definition 3 *Define decentralized overlapping fixed modes (DOFM) of $\mathscr{S}(\mathscr{K})$ as those modes of the system $\mathscr{S}$ which are fixed with respect to any dynamic LTI controller with the information flow structure $\mathscr{K}$.*

Theorem 4 states that the DOFMs of $\mathscr{S}(\mathscr{K})$ and the DFMs of $\mathscr{S}_\mu(\mathscr{K}_\mu)$, $\forall\, \mu \in \bar{l}$ are the same. Hence, the DOFMs of $\mathscr{S}(\mathscr{K})$ can be obtained from any of the systems $\mathscr{S}_1(\mathscr{K}_1), ..., \mathscr{S}_l(\mathscr{K}_l)$. The following procedure is used to determines the DOFMs of the system $\mathscr{S}(\mathscr{K})$ from the DFMs of the system $\mathscr{S}_\mu(\mathscr{K}_\mu)$, $\mu \in \bar{l}$.

Procedure 4 *Consider any arbitrary g belonging to $\bar{l}$. Let $sp(A)$ denote the set of eigenvalues of A. $\lambda \in sp(A)$ is a DOFM of the system $\mathscr{S}$ with respect to the information flow matrix $\mathscr{K}$, if there exists a permutation of $\{1, 2, ..., \nu_\mu\}$ denoted by the distinct integers $i_1, i_2, ..., i_{\nu_\mu}$, such that the rank of the matrix:*

$$\begin{bmatrix} A - \lambda I_n & \mathbf{B}_{i_1}^\mu & \mathbf{B}_{i_2}^\mu & \cdots & \mathbf{B}_{i_q}^\mu \\ \mathbf{C}_{i_{q+1}}^\mu & \mathbf{D}_{i_{q+1},i_1}^\mu & \mathbf{D}_{i_{q+1},i_2}^\mu & \cdots & \mathbf{D}_{i_{q+1},i_q}^\mu \\ \mathbf{C}_{i_{q+2}}^\mu & \mathbf{D}_{i_{q+2},i_1}^\mu & \mathbf{D}_{i_{q+2},i_2}^\mu & \cdots & \mathbf{D}_{i_{q+2},i_q}^\mu \\ \vdots & \vdots & \vdots & \ddots & \vdots \\ \mathbf{C}_{i_{\nu_\mu}}^\mu & \mathbf{D}_{i_{\nu_\mu},i_1}^\mu & \mathbf{D}_{i_{\nu_\mu},i_2}^\mu & \cdots & \mathbf{D}_{i_{\nu_\mu},i_q}^\mu \end{bmatrix} \tag{4.23}$$

is less than n for some $\mu \in \{0, 1, ..., \nu_\mu\}$.

Remark 3 *According to Procedure 4, the rank of a set of matrices given in (4.23) should be checked to find out if any of the eigenvalues of the matrix A is a DOFM of* *the system $\mathscr{S}(\mathscr{K})$. It can be easily verified that the number of these matrices grows exponentially by ν_μ (the number of complete bipartite subgraphs of $\mathscr{G}_\mu$). Therefore, in order to reduce the required computations, it is rather desirable to choose the graph $\mathscr{G}_\mu$ from the set of graphs $\{\mathscr{G}_1, ..., \mathscr{G}_l\}$, such that it has the minimum number of complete bipartite subgraphs. Moreover, if there is more than one such candidate, the one with fewer number of vertices is more preferable.*

Corollary 2 *The system $\mathscr{S}(\mathscr{K})$ is stabilizable by means of a dynamic LTI controller if and only if it does not have any DOFM in the closed right-half plane with respect to the information flow matrix $\mathscr{K}$.*

Proof. The proof follows immediately from Theorem 4. ■

The following theorem presents a method to characterize the DOFMs of the system $\mathscr{S}(\mathscr{K})$ in terms of the transmission zeros of a set of systems.

Theorem 5 *$\lambda \in sp(A)$ is a DOFM of the system $\mathscr{S}(\mathscr{K})$ if and only if it is a transmission zero of the following system:*

$$\dot{\mathbf{x}}(t) = A\mathbf{x}(t) + \begin{bmatrix} \mathbf{B}^l_{i_1} & \mathbf{B}^l_{i_2} & \cdots & \mathbf{B}^l_{i_q} \end{bmatrix} \mathbf{u}(t)$$
$$\mathbf{y}(t) = \begin{bmatrix} \mathbf{C}^l_{i_1} \\ \mathbf{C}^l_{i_2} \\ \vdots \\ \mathbf{C}^l_{i_q} \end{bmatrix} \mathbf{x}(t) + \begin{bmatrix} 0 & \mathbf{D}^l_{i_1 i_2} & \cdots & \mathbf{D}^l_{i_1 i_q} \\ \mathbf{D}^l_{i_2 i_1} & 0 & \cdots & \mathbf{D}^l_{i_2 i_q} \\ \vdots & \vdots & \ddots & \vdots \\ \mathbf{D}^l_{i_q i_1} & \mathbf{D}^l_{i_q i_2} & \cdots & 0 \end{bmatrix} \mathbf{u}(t) \tag{4.24}$$

for any $q \in \{1, ..., \nu_l\}$ and any arbitrary subset $\{i_1, i_2, ..., i_q\}$ of $\{1, 2, ..., \nu_l\}$.

Proof. It was shown earlier that the DOFMs of the system $\mathscr{S}(\mathscr{K})$ are the same as the DFMs of the system $\mathscr{S}_l(\mathscr{K}_l)$. Furthermore, since the matrix $\mathscr{K}_l$ is diagonal (because the graph $\mathscr{G}_l$ is composed of some disjoint edges), it results from [2] that the DFMs of the system $\mathscr{S}_l(\mathscr{K}_l)$ are the same as the common transmission zeros of the systems given by (4.24). This completes the proof. ■

Remark 4 *The results of Theorem 5 are obtained in [2] for the particular case when the information flow matrix $\mathscr{K}$ is block diagonal. Furthermore, the system given by (4.24) is constructed by using Kronecker product in [2], while it is formed by means of graph theory in this chapter. Therefore, Theorem 5 presents the results for the most general information flow structure compared to the ones given in [2].*

4.4.2 Optimal LTI controller

Assume that the system $\mathscr{S}$ is stabilizable with respect to the information flow matrix $\mathscr{K}$ by means of a dynamic LTI controller. It is desired to find a LTI controller with the zero initial state and the transfer function matrix $K(s)$ corresponding to the information flow structure $\mathscr{K}$, such that it minimizes the following LQR performance index:

$$J := \int_0^\infty \left(x(t)^T Q x(t) + u(t)^T R u(t)\right) dt \tag{4.25}$$

where $R \in \Re^{m\times m}$ and $Q \in \Re^{n\times n}$ are positive definite and positive semi-definite matrices, respectively, and where:

$$u(t) = \left[\, u_1(t)^T \; u_2(t)^T \; \cdots \; u_\nu(t)^T \,\right]^T \tag{4.26}$$

Lemma 1 *The matrix Φ_1 corresponding to the information flow matrix $\mathscr{K}_1$ is equal to I_m.*

Proof. : The proof follows directly from the procedure of constructing the graph $\mathscr{G}_1$ and Theorem 2. ∎

The transfer function matrix $K_1(s)$ constructed in terms of $K(s)$ in the proof of Theorem 3 (for $\mu = 1$) will be used in the next Theorem.

Theorem 6 *Consider the systems $\mathscr{S}$ and $\mathscr{S}_1$ under the controllers $K(s)$ and $K_1(s)$, respectively. Assume that the matrix $I_r - DK(s)$ is nonsingular. Then $J = J_1$, where:*

$$J_1 := \int_0^\infty \left(\mathbf{x}_1(t)^T Q \mathbf{x}_1(t) + \mathbf{u}_1(t)^T R \mathbf{u}_1(t)\right) dt \tag{4.27}$$

Proof. It follows from the proof of Theorem 3 (with $\mu = 1$) that $x(t) = \mathbf{x}_1(t)$ for all $t \geq 0$. Besides, it results from this equality and $\Phi_1 = I_m$ (Lemma 1), that $u(t) = \mathbf{u}_1(t)$ for all $t \geq 0$. This completes the proof. ∎

Theorem 6 states that in order to find a controller $K(s)$ which minimizes the performance index (4.25) for the system $\mathscr{S}$ while it meets the information flow constraint given by $\mathscr{K}$, one can equivalently pursue the following two steps:

1. Design the decentralized LTI controller $K_1(s)$ in such a way that it minimizes the performance index (4.27) for the system $\mathscr{S}_1(\mathscr{K}_1)$.
2. Find the controller $K(s)$ from the relation $K(s) = \Phi_1 K_1(s) \bar{\Phi}_1$.

It is to be noted that the decentralized optimal control design problem has been studied intensively in the literature, and a number of approaches for obtaining an optimal or a near-optimal decentralized controller are given accordingly, e.g., see [15, 16, 17].

4.5 Non-LTI control law

In this section, the procedure of designing different types of controllers, such as periodic or sampled-data control laws, for the system $\mathscr{S}(\mathscr{K})$ is investigated. Moreover, a necessary and sufficient condition for the stabilizability of the system $\mathscr{S}(\mathscr{K})$ is given. To develop the remaining results of this work, it is hereafter assumed that the system $\mathscr{S}$ is strictly proper, i.e. $D = 0$.

4.5.1 Generalized sampled-data hold function

Periodic control design using generalized sampled-data hold function (GSHF) and its advantages have been studied intensively in the literature [18, 19, 20]. Assume that it is desired to design a GSHF for the system $\mathscr{S}$, which complies with the information flow structure $\mathscr{K}$. Let this GSHF be denoted by $F(t)$. Hence, the hold controller will be as follows:

$$u(t) = F(t)y[\kappa], \quad \kappa h \leq t < (\kappa+1)h, \quad \kappa \geq 0 \tag{4.28}$$

where h represents the sampling periodic. Note that the discrete argument corresponding to the samples of any signal is enclosed in brackets (e.g., $y[\kappa] := y(\kappa h)$).

Theorem 7 *The systems $\mathscr{S}(\mathscr{K}), \mathscr{S}_1(\mathscr{K}_1), ..., \mathscr{S}_l(\mathscr{K}_l)$ are all* analogous *with respect to the set of all hold controllers (GSHFs).*

Proof. To prove the theorem, it suffices to show that $\mathscr{S}(\mathscr{K})$ and $\mathscr{S}_\mu(\mathscr{K}_\mu)$ are *analogous* with respect to all hold controllers, for any $\mu \in \bar{l}$. Consider a GSHF $F(t)$ which complies with the information flow structure $\mathscr{K}$. Utilize the proper transformation on $F(t)$ to obtain the equivalent hold function $F_\mu(t)$ for the system $\mathscr{S}_\mu(\mathscr{K}_\mu)$. Note that $F_\mu(t)$ can be attained using the mapping between the components of $\mathbf{K}$ and $\mathbf{K}_\mu$ (see Remark 1). Since $F(t)$ and $F_\mu(t)$ comply with the information flow matrices $\mathscr{K}$ and $\mathscr{K}_\mu$, respectively, it is straightforward to show that $F(t) = \Phi_\mu F_\mu(t)\bar{\Phi}_\mu$. On the other hand, it follows from (4.28) that:

$$\dot{x}(t) = Ax(t) + BF(t)Cx[\kappa] \tag{4.29}$$

and consequently:

$$\begin{aligned} \dot{\mathbf{x}}_\mu(t) &= \mathbf{A}\mathbf{x}_\mu(t) + \mathbf{B}^\mu F_\mu(t)\mathbf{C}^\mu \mathbf{x}_\mu[\kappa] \\ &= \mathbf{A}\mathbf{x}_\mu(t) + B\Phi_\mu F_\mu(t)\bar{\Phi}_\mu C\mathbf{x}_\mu[\kappa] \\ &= \mathbf{A}\mathbf{x}_\mu(t) + BF(t)C\mathbf{x}_\mu[\kappa] \end{aligned} \tag{4.30}$$

for all $t \in [\kappa h, (\kappa+1)h)$, $\kappa \geq 0$. The equations (4.29) and (4.30), and the equality $x(0) = \mathbf{x}_\mu(0)$ result in the relation $x(t) = \mathbf{x}_\mu(t)$ for all $t \geq 0$. Conversely, for any GSHF $F_\mu(t)$ complying with the information flow matrix $\mathscr{K}_\mu$, it is straightforward to show that the state of the system $\mathscr{S}$ under the GSHF $F(t) = \Phi_\mu F_\mu(t)\bar{\Phi}_\mu$ is identical to that of the system $\mathscr{S}_\mu$ under $F_\mu(t)$. ■

Theorem 7 states that the problem of designing a GSHF for the system $\mathscr{S}(\mathscr{K})$ can be formulated as the problem of designing a GSHF for the system $\mathscr{S}_\mu(\mathscr{K}_\mu)$ for any $\mu \in \bar{l}$. However, due to the decentralized structure of the control for $\mathscr{S}_\mu(\mathscr{K}_\mu)$, $\mu \in \bar{l}$, the corresponding GSHF design can be accomplished by using the existing methods [21, 22].

4.5.2 Sampled-data controller

A typical sampled-data controller consists of a sampler, a zero-order hold (ZOH) and a discrete-time controller. It is to be noted that a sampled-data controller acts as a time-varying control law for the continuous-time system. It is desired in this subsection to present a method for designing a sampled-data controller for the system $\mathscr{S}$, whose structure complies with a given information flow matrix $\mathscr{K}$. Throughout the remainder of this chapter, the term *linear shift-invariant* (LSI) will be used instead of LTI, for discrete-time systems.

Theorem 8 *The systems $\mathscr{S}(\mathscr{K}), \mathscr{S}_1(\mathscr{K}_1), ..., \mathscr{S}_l(\mathscr{K}_l)$ are all* analogous *with respect to the set of all LSI sampled-data controllers.*

Proof. Denote the sampling period with h, and the discrete-time equivalent models of the systems $\mathscr{S}, \mathscr{S}_1, ..., \mathscr{S}_l$ with $\bar{\mathscr{S}}, \bar{\mathscr{S}}_1, ..., \bar{\mathscr{S}}_l$, respectively. Assume that the system $\bar{\mathscr{S}}$ is represented by:

$$\begin{aligned} x[\kappa+1] &= \bar{A}x[\kappa] + \bar{B}u[\kappa] \\ y[\kappa] &= Cx[\kappa] \end{aligned} \tag{4.31}$$

Similarly, let the system $\bar{\mathscr{S}}_\mu$ be represented by:

$$\begin{aligned} \mathbf{x}_\mu[\kappa+1] &= \bar{A}\mathbf{x}_\mu[\kappa] + \bar{\mathbf{B}}^\mu \mathbf{u}_\mu[\kappa] \\ \mathbf{y}_\mu[\kappa] &= \mathbf{C}^\mu \mathbf{x}_\mu[\kappa], \qquad \mu \in \bar{l} \end{aligned} \tag{4.32}$$

It can be easily verified that:

$$\bar{\mathbf{B}}^\mu = \int_0^h e^{\tau A}\mathbf{B}^\mu \, d\tau = \int_0^h e^{\tau A} B \, d\tau \times \Phi_\mu = \bar{B}\Phi_\mu \tag{4.33}$$

It results from (4.31), (4.32), and (4.33) that the state-space matrices of $\bar{\mathscr{S}}$ are related to those of $\bar{\mathscr{S}}_\mu$, exactly the same way the state-space matrices of $\mathscr{S}$ and $\mathscr{S}_\mu$ are related. Hence, the systems $\bar{\mathscr{S}}$ and $\bar{\mathscr{S}}_\mu$ are *analogous* with respect to the LSI controllers. Consider now a discrete-time LSI controller with the transfer function matrix $\bar{K}(z)$ for the system $\bar{\mathscr{S}}(\mathscr{K})$. Construct a discrete-time LSI controller with the transfer function matrix $\bar{K}_\mu(z)$ for the system $\bar{\mathscr{S}}_\mu(\mathscr{K}_\mu)$, such that it corresponds to the controller $\bar{K}(z)$ for $\bar{\mathscr{S}}(\mathscr{K})$. This controller can be obtained from the mapping between the components of $\mathbf{K}$ and $\mathbf{K}_\mu$. It is straightforward to show that $\bar{K}(z) = \Phi_\mu \bar{K}_\mu(z)\bar{\Phi}_\mu$. Applying the controller $\bar{K}(z)$ to the system $\bar{\mathscr{S}}$ and the controller $\bar{K}_\mu(z)$ to $\bar{\mathscr{S}}_\mu$, one can conclude (using an approach similar to the one given in the proof of Theorem 3) that $x[\kappa] = \mathbf{x}_\mu[\kappa]$ and $u[\kappa] = \Phi_\mu \mathbf{u}_\mu[\kappa]$ for any $\kappa \geq 0$. Therefore,

$$\begin{aligned} x(t) &= e^{(t-\kappa h)A} x[\kappa] + \int_{\kappa h}^{t} e^{(\tau-\kappa h)A} Bu[\kappa]\, d\tau \\ &= e^{(t-\kappa h)A} \mathbf{x}_\mu[\kappa] + \int_{\kappa h}^{t} e^{(\tau-\kappa h)A} B\Phi_\mu \mathbf{u}_\mu[\kappa]\, d\tau \\ &= e^{(t-\kappa h)A} \mathbf{x}_\mu[\kappa] + \int_{\kappa h}^{t} e^{(\tau-\kappa h)A} \mathbf{B}^\mu \mathbf{u}_\mu[\kappa]\, d\tau \\ &= \mathbf{x}_\mu(t) \end{aligned} \tag{4.34}$$

for any $t \in [\kappa h, (\kappa+1)h)$, $k \geq 0$. Similarly, it can be easily verified that given any controller $\bar{K}_\mu(z)$ for the system $\mathscr{S}_\mu(\mathscr{K})$, the controller $\bar{K}(z) := \Phi_\mu \bar{K}_\mu(z) \bar{\Phi}_\mu$ corresponds to the information flow matrix $\mathscr{K}$. Moreover, the state of the system $\mathscr{S}$ under the controller $\bar{K}(z)$ is the same as that of $\mathscr{S}_\mu$ under $\bar{K}_\mu(z)$. ■

It is assumed in the proof of Theorem 8 that $D = 0$. However, its results can be easily extended to the case when $D \neq 0$. Note that finding a sampled-data decentralized control law to achieve certain design objectives has been investigated in the literature, e.g, see [23].

4.5.3 Finite-dimensional linear time-varying controller

It is well-known that finite-dimensional linear time-varying (LTV) controllers are superior to their LTI counterparts in many control applications [21]. It is desired in this subsection to present a procedure for designing a finite-dimensional LTV controller complying with the information flow matrix $\mathscr{K}$, for the system $\mathscr{S}$. Note that throughout this work, the term "finite-dimensional LTV controller" refers to a control law which can be represented by the following state-space model:

$$\begin{aligned} \dot{\tilde{x}}(t) &= \tilde{A}(t)\tilde{x}(t) + \tilde{B}(t)\tilde{u}(t) \\ \tilde{y}(t) &= \tilde{C}(t)\tilde{x}(t) + \tilde{D}(t)\tilde{u}(t) \end{aligned} \tag{4.35}$$

Theorem 9 *The systems $\mathscr{S}(\mathscr{K}), \mathscr{S}_1(\mathscr{K}_1), ..., \mathscr{S}_l(\mathscr{K}_l)$ are all* analogous *with respect to the set of all finite-dimensional LTV controllers.*

The foregoing theorem extends the results of Theorem 3 to the case when the controllers are finite-dimensional LTV (as opposed to LTI). The proof of Theorem 9 is similar to that of Theorem 3 (but should be carried out in the time-domain). The details of the proof are omitted here. However, the statement that $\mathscr{S}(\mathscr{K})$ and $\mathscr{S}_\mu(\mathscr{K}_\mu)$, $\mu \in \bar{l}$, are *analogous* with respect to all finite-dimensional LTV controllers can be intuitively justified as follows:

One can easily verify by using the comments given in the proofs of Theorems 1 and 2, that $\mathbf{B}^\mu$ is derived from B by rearranging its columns and repeating some of them (repetition results from the fact that some of the vertices in set 1 of the graph

$\mathscr{G}$ have recurred to construct the graph $\mathscr{G}_\mu$). Analogously, $\mathbf{C}^\mu$ is derived from C by rearranging its rows and repeating some of them. These repetitions and rearrangements and their interpretations are described below:

1. *Repetition of the rows of C indicates that some of the outputs of $\mathscr{S}$ are duplicated to construct the system $\mathscr{S}_\mu$. To justify the necessity of this recurrence, assume that one output of the system $\mathscr{S}$ contributes to two different control inputs. This means that the corresponding control agent is not localized, and hence the corresponding information flow structure is not decentralized. However, by duplicating this output of the system to create a redundant output, and by applying the two resulting outputs to the two above-mentioned control inputs, the resultant control structure will be decentralized, while its functionality is essentially equivalent to the original control system.*
2. *Regarding the repetition in the columns of $\mathbf{B}^\mu$, assume that two outputs of the system contribute to one control agent. Since the controller is linear, one can split the control agent to two sub-agents such that each of the two outputs of the system goes to one of these sub-agents. The control signal of the original control agent is, in fact, equal to the summation of the control signals of these two sub-agents (this results from the principle of superposition). Again, the functionality of the resultant control system is equivalent to the original one, while its structure is decentralized.*
3. *The rearrangement of the rows and the columns of C and B is equivalent to the reordering of the inputs and the outputs of $\mathscr{S}$, and has no impact on the operation of the overall control system.*

Taking the aforementioned interpretations into consideration, the system $\mathscr{S}_\mu$ is indeed constructed from $\mathscr{S}$ in such a way that the control structure $\mathscr{K}$ is converted to a decentralized structure $\mathscr{K}_\mu$, while essentially both control systems perform identically.

Theorem 9 implies that to design a finite-dimensional LTV controller for the system $\mathscr{S}(\mathscr{K})$, one can first design a LTV controller for one of the systems $\mathscr{S}_1(\mathscr{K}_1),...,\mathscr{S}_l(\mathscr{K}_l)$. This result will be exploited in the following section to present one of the main contributions of the present work.

4.5.4 General controller

The objective of this subsection is to find out under what conditions the system $\mathscr{S}(\mathscr{K})$ is stabilizable by means of a general control law (i.e. nonlinear and time-varying), when there exists no stabilizing LTI controller.

Theorem 10 *The systems $\mathscr{S}(\mathscr{K})$ and $\mathscr{S}_1(\mathscr{K}_1)$ are* analogous *with respect to any type of controller (i.e. nonlinear or time-varying).*

Proof. As pointed out in the discussion following Theorem 9, the configurations of the systems $\mathscr{S}$ and $\mathscr{S}_1$ are essentially equivalent. In other words, the system $\mathscr{S}_1$

is obtained from $\mathscr{S}$ by introducing some redundant outputs or control agents and reordering them, in such a way that the information flow structure $\mathscr{K}$ is converted to $\mathscr{K}_1$. Note that according to Lemma 1, $B = \mathbf{B}^1$. Hence, the state of the closed-loop system corresponding to the pair $(\mathscr{S}, \mathscr{K})$ is identical to that of the pair $(\mathscr{S}_1, \mathscr{K}_1)$, regardless of the type of the control law. ■

It is to be noted that unlike $\mathscr{S}(\mathscr{K})$ and $\mathscr{S}_1(\mathscr{K}_1)$, the systems $\mathscr{S}(\mathscr{K})$ and $\mathscr{S}_\mu(\mathscr{K}_\mu)$, $\mu \in \{2,3,...,l\}$, are not *analogous* with respect to any type of controller, in general. This results from the fact that the superposition principle presented in item 2 of the discussion following Theorem 9 does not apply here, as the controllers are nonlinear.

Remark 5 *It follows immediately from Theorem 10 that the system $\mathscr{S}(\mathscr{K})$ is stabilizable if and only if the system $\mathscr{S}_1(\mathscr{K}_1)$ is stabilizable.*

It is shown in [3] that a system is stabilizable with respect to a block-diagonal information flow matrix (i.e. decentralized control structure) if and only if the system does not any unstable quotient fixed mode (QFM). However, QFM is only defined for decentralized control structures. In the following, this notion is extended to the general information flow structure and its property is investigated accordingly.

Definition 4 *$\lambda \in sp(A)$ is a quotient overlapping fixed mode (QOFM) of the system $\mathscr{S}$ with respect to the information flow matrix $\mathscr{K}$, if λ cannot be eliminated by using any type of controller complying with the structure of $\mathscr{K}$.*

Theorem 11 *The QOFMs of the system $\mathscr{S}(\mathscr{K})$ are the same as the QFMs of the system $\mathscr{S}_\mu(\mathscr{K}_\mu)$, $\forall \mu \in \bar{l}$.*

Proof. It follows from Theorem 10 that the QOFMs of the system $\mathscr{S}(\mathscr{K})$ are the same as the QFMs of the system $\mathscr{S}_1(\mathscr{K}_1)$. To complete the proof, it suffices to show that the QFMs of the system $\mathscr{S}_1(\mathscr{K}_1)$ are the same as those of the system $\mathscr{S}_\mu(\mathscr{K}_\mu)$, for $\mu = 2,3,...,l$. This can be deduced from the following argument:

- The systems $\mathscr{S}_1(\mathscr{K}_1),...,\mathscr{S}_l(\mathscr{K}_l)$ all have the same A-matrix, and hence the same modes.
- It is shown in [3, 24] that all of the non-QFMs of any system can be eliminated by using a proper finite-dimensional LTV controller.
- Theorem 9 states that the systems $\mathscr{S}_1(\mathscr{K}_1),...,\mathscr{S}_l(\mathscr{K}_l)$ are all *analogous* to each other with respect to finite-dimensional LTV controllers.

■

Corollary 3 *The system $\mathscr{S}(\mathscr{K})$ is stabilizable if and only if it does not have any unstable QOFM.*

Proof. The proof follows immediately from Remark 5 and Theorem 11. ■

4.6 Comparison with existing methods

4.6.1 Comparison with the work presented in [8]

Consider the system $\mathscr{S}$ with the following state-space representation:

$$\dot{x}(t) = Ax + B_1u_1(t) + B_2u_2(t) \tag{4.36}$$

where $A \in \Re^{3\zeta\times 3\zeta}$, and

$$B_1 = \begin{bmatrix} B_{11} \\ 0_{\zeta\times\zeta_1} \\ 0_{\zeta\times\zeta_1} \end{bmatrix},\ B_2 = \begin{bmatrix} 0_{\zeta\times\zeta_2} \\ 0_{\zeta\times\zeta_2} \\ B_{32} \end{bmatrix},\ x(t) = \begin{bmatrix} x_1(t) \\ x_2(t) \\ x_3(t) \end{bmatrix} \tag{4.37}$$

and $x_i(t) \in \Re^{\zeta}$, $i = 1,2,3$. The outputs of this system are assumed to be the same as its state variables. It is desired now to design a stabilizing structurally constrained static controller for $\mathscr{S}$ with the information flow matrix :

$$\mathscr{K} = \begin{bmatrix} 1\ 1\ 0 \\ 0\ 1\ 1 \end{bmatrix} \tag{4.38}$$

This decentralized overlapping problem is investigated in [8], where the expansion approach is used to solve the problem (see [4] for a numerical version of this example). In this method, the system $\mathscr{S}$ is converted to another system, which is referred to as the expanded system. Subsequently, it is stated that if the expanded system can be stabilized, then the system $\mathscr{S}$ is stabilizable as well. However, since the expanded system is inherently uncontrollable, this approach might be inefficient. It is desired now to demonstrate the effectiveness of the method proposed in this chapter for this example. Using the proposed method, one can easily conclude that the DOFMs of the system $\mathscr{S}$ consist of unobservable modes, uncontrollable modes, and any mode λ for which at least one of the following two matrices:

$$\begin{bmatrix} A-\lambda I_n & B_1 \\ H_1 & o_{2\zeta\times\zeta_1} \end{bmatrix},\ \begin{bmatrix} A-\lambda I_n & B_2 \\ H_2 & o_{2\zeta\times\zeta_2} \end{bmatrix} \tag{4.39}$$

loses rank, where $H_1 = \begin{bmatrix} I_{2\zeta} & o_{2\zeta\times\zeta} \end{bmatrix}$ and $H_2 = \begin{bmatrix} o_{2\zeta\times\zeta} & I_{2\zeta} \end{bmatrix}$. Assume now that the system does not have any unstable DOFM. One can use Procedures 1, 2 and 3 to obtain the matrix $\mathbf{K}_2$ as follows:

$$\mathbf{K}_2 = \begin{bmatrix} k_{11} & 0_{\zeta_1\times\zeta} & 0_{\zeta_1\times\zeta} \\ 0_{\zeta_1\times\zeta} & k_{12} & 0_{\zeta_1\times\zeta} \\ 0_{\zeta_2\times\zeta} & k_{22} & 0_{\zeta_2\times\zeta} \\ 0_{\zeta_2\times\zeta} & 0_{\zeta_2\times\zeta} & k_{23} \end{bmatrix} \tag{4.40}$$

which corresponds to the system $\mathscr{S}_2$ with the following state-space representation:

$$\dot{\mathbf{x}}_2(t) = A\mathbf{x}_2(t) + \mathbf{B}_1^2\mathbf{u}_1^2(t) + \mathbf{B}_2^2\mathbf{u}_2^2(t) + \mathbf{B}_3^2\mathbf{u}_3^2(t) \tag{4.41}$$

where:

$$\mathbf{x}_2(t) = \begin{bmatrix} \mathbf{x}_1^2(t) \\ \mathbf{x}_2^2(t) \\ \mathbf{x}_3^2(t) \end{bmatrix}, \quad \mathbf{B}_1^2 = B_1, \quad \mathbf{B}_2^2 = \begin{bmatrix} B_1 \ B_2 \end{bmatrix}, \quad \mathbf{B}_3^2 = B_2 \tag{4.42}$$

and $\mathbf{x}_i^2(t) \in \Re^{\varsigma}$, $i = 1,2,3$. It can be concluded from Corollary 1 that designing a *static* structurally constrained controller for the system $\mathscr{S}$ is identical to designing a *static* decentralized controller for the system $\mathscr{S}_2$ (i.e., $\mathbf{u}_i^2(t)$ is constructed in terms of $\mathbf{x}_i^2(t)$ for $i = 1,2,3$). The latter problem can be solved by using either the LMI method proposed in [8] (where this example is presented), or other existing methods, e.g. [15].

4.6.2 Comparison with the work presented in [25]

A method is proposed in [25] for strictly proper systems to determine whether the system is stabilizable with respect to a given information flow matrix by means of LTI controllers. However, this method has the following deficiencies compared to the present work:

1. It cannot be extended to the general proper systems.
2. It translates the stabilizability of a system by means of LTI controllers to that of another system which is, in fact, $\mathscr{S}_1$. However, this may require that the ranks of a huge number of matrices to be checked in order to find out whether the system is stabilizable. For instance, assume that the system $\mathscr{S}$ is composed of 100 SISO subsystems, and that the corresponding information flow matrix $\mathscr{K}$ is a 100×100 matrix, with the first entries of the odd rows and the last entries of the even rows all equal to zero, and the remaining entries all equal to 1. It is straightforward to show that the number of matrices whose ranks need to be checked by using the method given in [25] is equal to 2^{100}, while the system $\mathscr{S}_2$ can be constructed in such a way that the number of matrices whose ranks need to be checked is equal to 2^2. This sizable difference demonstrates the efficiency of the present work.

4.7 Numerical examples

Example 1 Consider the system $\mathscr{S}$ consisting of three SISO subsystems with the following state-space matrices:

$$A=\begin{bmatrix}1\,0\,1\\0\,1\,2\\1\,2\,3\end{bmatrix},\quad B=\begin{bmatrix}0\,1\,0\\0\,0\,1\\1\,0\,1\end{bmatrix},\quad C=\begin{bmatrix}1\,0\,0\\0\,0\,1\\1\,2\,2\end{bmatrix},\quad D=0_{3\times 3} \tag{4.43}$$

Consider the information flow matrix $\mathscr{K}=\begin{bmatrix}1\,0\,1\\0\,1\,1\\1\,1\,1\end{bmatrix}$, which corresponds to the following control structure:

$$\mathbf{K}=\begin{bmatrix}k_{11} & 0 & k_{13}\\0 & k_{22} & k_{23}\\k_{31} & k_{32} & k_{33}\end{bmatrix} \tag{4.44}$$

This is, in fact, a BAS (or BBD) controller [8, 9]. The following matrix $\mathbf{K}_2$ can be obtained using Procedures 1, 2 and 3:

$$\mathbf{K}_2=\begin{bmatrix}k_{22} & k_{23} & 0 & 0\\k_{32} & k_{33} & 0 & 0\\0 & 0 & k_{11} & 0\\0 & 0 & k_{31} & 0\\0 & 0 & 0 & k_{13}\end{bmatrix} \tag{4.45}$$

Note that for this particular example, $\mathscr{G}_2$ is the best candidate in terms of the subsequent computational complexity. The matrices Φ_2 and $\bar{\Phi}_2$ can be obtained from Theorem 2 as follows:

$$\Phi_1=\begin{bmatrix}0\,0\,1\,0\,1\\1\,0\,0\,0\,0\\0\,1\,0\,1\,0\end{bmatrix},\quad \Phi_2=\begin{bmatrix}0\,1\,0\\0\,0\,1\\1\,0\,0\\0\,0\,1\end{bmatrix} \tag{4.46}$$

Now, the system $\mathscr{S}_2$ can be easily constructed by using the equations (4.15), (4.16) and (4.46). It is desired now to design a structurally constrained controller $K(s)$ for the system $\mathscr{S}$ to achieve a settling time of 4 seconds. and an overshoot of less than 4.5%. It can be easily verified that these design specifications will be met by placing the dominant poles of the closed-loop system at $-1\pm 1i$. From Procedure 4, it is known that the system $\mathscr{S}$ does not have any DOFMs with respect to the information flow matrix $\mathscr{K}$. Now, using any decentralized pole placement method, e.g., the one proposed in [2] or [14], one can place the dominant poles of the closed-loop system $\mathscr{S}_2$ at $-1\pm 1i$, as discussed in Remark 2. For instance, using the method given in [2], the following control transfer functions are obtained:

$$
\begin{aligned}
&K_{11}(s)=K_{13}(s)=K_{31}(s)=1\\
&K_{22}(s)=\left(-89900-96100s-34100s^2-5480s^3-409s^4-11.5s^5\right)/\mathrm{Den}(s)\\
&K_{23}(s)=\left(-15700-20500s-8810s^2-1730s^3-160s^4-5.69s^5\right)/\mathrm{Den}(s)\\
&K_{32}(s)=\left(-64500-52500s-16900s^2-2740s^3-220s^4-7.05s^5\right)/\mathrm{Den}(s)\\
&K_{33}(s)=\left(-88000-64500s-19200s^2-2880s^3-219s^4-6.7s^5\right)/\mathrm{Den}(s)
\end{aligned}
\tag{4.47}
$$

where:

$$
\mathrm{Den}(s)=0.18s^6+9.95s^5+210.44s^4+2269.3s^3+13396s^2+41488s^1+53000 \tag{4.48}
$$

(the transfer function of the control component k_{ij} is represented by $K_{ij}(s)$). It is to be noted that using the above control law, the other poles of the closed-loop system will be located at $-4, -6, -7$ and -8.

Example 2 Consider the system $\mathscr{S}$ consisting of four SISO subsystems with the following state-space matrices:

$$
A=\begin{bmatrix}1&0&0&0&0&0\\0&-2&0&0&0&0\\0&0&-3&0&0&0\\1&1&1&1&0&0\\1&1&1&0&-2&0\\1&1&1&0&0&-3\end{bmatrix},\ B=\begin{bmatrix}1&0&0&0\\0&3&0&0\\0&1&0&0\\0&0&1&0\\0&0&0&3\\0&0&0&1\end{bmatrix},\ C=\begin{bmatrix}0&1&0&0\\0&0&1&0\\0&1&-4&0\\1&0&0&0\\0&0&0&1\\1&0&0&-4\end{bmatrix}^T \tag{4.49}
$$

and $D=0_{4\times4}$. Assume that the information flow matrix for this system is given as follows:

$$
\mathscr{K}=\begin{bmatrix}0&0&1&0\\0&1&0&1\\0&0&0&1\\1&0&0&0\end{bmatrix} \tag{4.50}
$$

One can find the matrices Φ_1 and $\bar{\Phi}_1$ (using Procedures 1, 2 and 3, and Theorem 2), from which one yields that the system $\mathscr{S}$ has two identical DOFMs at $\lambda=+1$ with respect to the information flow matrix $\mathscr{K}$ given by (4.50). Therefore, this system cannot be stabilized by means of a structurally constrained LTI controller. On the other hand, it can be easily verified by using the system $\mathscr{S}_1$ that $\mathscr{S}$ does not have any QOFMs. Hence, this system can be stabilized by means of a constrained LTV controller. Using the method given in [23], one can design a constrained stabilizing sampled-data controller for the system $\mathscr{S}(\mathscr{K})$. Consider a sampling period h equal to 1. The components of the controller will be as follows:

$$
\begin{aligned}
&\bar{K}_{22}(z)=\bar{K}_{34}(z)=0,\quad \bar{K}_{13}(z)=\bar{K}_{24}(z)=1,\\
&\bar{K}_{41}(z)=\left(3945z^5-8674z^4+1388z^3+116.2z^2-12.8z-1.139\right)\\
&\qquad\times\left(z^6+2.758z^5+877.1z^4-1822z^3+87.78z^2+24.71z+0.1927\right)^{-1}
\end{aligned}
\tag{4.51}
$$

where $\bar{K}_{ij}(z)$ represents the transfer function of the discrete-time LSI controller corresponding to k_{ij}.

4.8 Summary

This work tackles the control design problem for systems with constrained control structure. It is shown that certain modes of the system can be placed freely by means of a linear time-invariant (LTI) structurally constrained controller. The notion of a decentralized overlapping fixed mode (DOFM) is introduced to classify such modes, and an analytical method is given to identify them. In addition, it is shown that the system is stabilizable by means of a LTI structurally constrained controller, if and only if it does not have any unstable DOFM. Furthermore, a graph-theoretic algorithm is proposed to convert the structurally constrained control design problem (e.g. pole placement, optimal feedback, etc.) to the conventional decentralized control design problem. Design procedures for different types of controllers, such as periodic and sampled-data control laws, are also investigated. The notion of a quotient overlapping fixed mode (QOFM) is then introduced to determine whether the system can be stabilized by means of general (nonlinear and time-varying) structurally constrained controllers. It is shown that a system with no unstable QOFM can be stabilized by utilizing a finite-dimensional linear time-varying control law.

References

1. S. H. Wang and E. J. Davison, "On the stabilization of decentralized control systems," *IEEE Transactions on Automatic Control*, vol. 18, no. 5, pp. 473-478, 1973.
2. E. J. Davison and T. N. Chang, "Decentralized stabilization and pole assignment for general proper systems," *IEEE Transactions on Automatic Control*, vol. 35, no. 6, pp. 652-664, 1990.
3. Z. Gong and M. Aldeen, "Stabilization of decentralized control systems," *Journal of Mathematical Systems, Estimation, and Control*, vol. 7, no. 1, pp. 1-16, 1997.
4. A. I. Zecevic and D. D. Šiljak, "A new approach to control design with overlapping information structure constraint," *Automatica*, vol. 41, no. 2, pp. 265-272, 2005.
5. L. Bakule, J. Rodellar, and J. M. Rossell, "Contractibility of dynamic LTI controllers using complementary matrices," *IEEE Transactions on Automatic Control*, vol. 48, no. 7, pp. 1269-1274, 2003.
6. A. Iftar, "Overlapping decentralized dynamic optimal control," *International Journal of Control.* vol. 58, no. 1, pp. 187-209, 1993.
7. A. Iftar, "Decentralized optimal control with overlapping decompositions," *IEEE International Conference on Systems Engineering*, pp. 299-302, 1991.
8. D. D. Šiljak and A. I. Zecevic, "Control of large-scale systems: Beyond decentralized feedback," *Annual Reviews in Control*, vol. 29, no. 2, pp. 169-179, 2005.
9. P. P. Groumpos, "Structural modelling and optimisation of large scale systems," *IEE Control Theory and Applications*, vol. 141, no. 1, pp. 1-11, 1994.
10. A. P. Leros and P. P. Groumpos, "The time-invariant BAS decentralized large-scale linear regulator proble," *International Journal of Control*, vol. 46, no. 1, pp. 129-152, 1987.

11. S. S. Stankovic and M. J. Stanojevic, and D. D. Šiljak, "Decentralized overlapping control of a platoon of vehicles," *IEEE Transactions on Control Systems Technology*, vol. 8, no. 5, pp. 816-832, 2000.
12. Y. Ebihara and T. Hagiwara, "Structured controller synthesis using LMI and alternating projection method," in *Proceedings of 42nd IEEE Conference on Decision and Control*, pp. 5632-5637, 2003.
13. J. Han and R. E. Skelton, "An LMI optimization approach for structured linear controllers," in *Proceedings of 42nd IEEE Conference on Decision and Control*, pp. 5143-5148, 2003.
14. M. S. Ravi, J. Rosenthal, and X. A. Wang, "On decentralized dynamic pole placement and feedback stabilization," *IEEE Transactions on Automatic Control*, vol. 40, no. 9, pp. 1603-1614, 1995.
15. D. D. Šiljak, Decentralized control of complex systems, Cambridge: Academic Press, 1991.
16. K. D. Young, "On near optimal decentralized control," *Automatica*, vol. 21, no. 5, pp. 607-610, 1985.
17. H. T. Toivoneh and P. M. Makila, "A descent anderson-moore algorithm for optimal decentralized control," *Automatica*, vol. 21, no. 6, pp. 743-744, 1985.
18. M. Rossi and D. E. Miller, "Gain/phase margin improvement using static generalized sampled-data hold functions," *Systems & Control Letters*, vol. 37, no. 3, pp. 163-172, 1999.
19. P. T. Kabamba, "Control of linear systems using generalized sampled-data hold functions," *IEEE Transactions on Automatic Control*, vol. 32, no. 9, pp. 772-783, 1987.
20. J. L. Yanesi and A. G. Aghdam, "Optimal generalized sampled-data hold functions with a constrained structure," in *Proceedings of 2006 American Control Conference*, Minneapolis, Minnesota, 2006.
21. A. G. Aghdam, "Decentralized control design using piecewise constant hold functions," in *Proceedings of 2006 American Control Conference*, Minneapolis, Minnesota, 2006.
22. S. H. Wang, "Stabilization of decentralized control systems via time-varying controllers," *IEEE Transactions on Automatic Control*, vol. 27, no. 3, pp. 741-744, 1982.
23. Ü. Özgüner and E. J. Davison, "Sampling and decentralized fixed modes," *Proceedings of the 1985 American Control Conference*, pp. 257-262, 1985.
24. B. Anderson and J. Moore, "Time-varying feedback laws for decentralized control," *IEEE Transactions on Automatic Control*, vol. 26, no. 5, pp. 1133-1139, 1981.
25. V. Pichai, M. E. Sezer, and D. D. Šiljak, "A graph-theoretic characterization of structurally fixed modes," *Automatica*, vol. 20, no. 2, pp. 247-250, 1984.

Chapter 5
Elimination of Decentralized Fixed Modes via Optimal Information Exchange

5.1 Introduction

Numerous real-world systems can be modeled as the interconnected systems consisting of a number of subsystems. The control of an interconnected system is often carried out by means of a set of local controllers, corresponding to the interacting subsystems [1, 2]. It is sometimes assumed that the local controllers can fully communicate with each other in order to elevate their effectiveness over the entire system cooperatively. However, this design technique is often problematic as the required data transmission between two particular local controllers (or equivalently, two subsystems) can be unjustifiably expensive or occasionally infeasible. Consequently, it is normally desired that the local controllers either exchange partial information or act independently of each other. The latter case, where the overall controller consists of a set of isolated local controllers, is referred to as decentralized control in the literature [3, 4, 5]. The control structure in a decentralized control system is, in fact, block-diagonal. It is to be noted that the decentralized control theory has found applications in large space structures, power systems, communication networks, etc. [6, 7, 8, 9]. A wide variety of properties of the decentralized control systems are extensively studied in the literature and different design techniques are proposed [10, 11, 12, 13].

One of the important problems in decentralized control design for interconnected systems is the stabilizability verification. The notion of a decentralized fixed mode (DFM) was introduced in [14] to identify those modes of a system which are fixed with respect to any LTI decentralized controller. Various methods are introduced in the literature to characterize DFMs [4, 15, 16, 17, 18]. For instance, the method given in [4] provides the existence conditions for DFMs in terms of the rank of a set of matrices. As a computationally more efficient technique, the paper [18] proposes a simple graph-theoretic approach to verify whether an unrepeated mode of the system is a DFM. It is to be noted that a controllable and observable system may have a DFM, due to the isolation of its local controllers.

Given an interconnected system with at least one unstable DFM, the question arises: Can a stabilizing LTI controller be designed for this system by establishing new information flow channels in the control configuration (which will roughly possess a decentralized structure)? This question has been addressed in a number of papers to some extent by making certain permissible interactions between the local controllers. The work [19] uses this idea to tackle the underlying problem, but it fails to obtain the minimum number of required interactions to achieve stabilizability. This shortcoming limits the effectiveness of the method in practical applications considerably. The paper [20] deals with the pole-assignability problem for interconnected systems by means of partially interacting LTI local controllers. A cost is first attributed to the communication link between any pair of local controllers in order to formulate the implementation expenditure. Nevertheless, the work [20] considers only a particular class of the modes, due to the complexity of the problem in the general case. This particular class is, in fact, the fixed modes which result from the structure of the system, rather than an exact matching of the parameters of the system. This class of fixed modes is referred to as structurally fixed mode (SDFM) [21]. The method proposed in [20] leads to a near-optimal solution by solving two separate optimization problems. A simpler method to handle the same problem (i.e., eliminating the SDFMs of a system) is more recently presented in [22].

When some local controllers are capable of interacting with each other, the overall controller is said to be a decentralized overlapping controller [23]. The stabilizability of an interconnected system by means of LTI decentralized overlapping controllers has been investigated thoroughly in [23] and [24] using the new notions of decentralized overlapping fixed mode (DOFM) and quotient overlapping fixed mode (QOFM). It is to be noted that the decentralized overlapping control theory was initially introduced for the systems with some overlapping subsystems. For this type of systems, it is often desired to design decentralized overlapping controllers whose overlapping structure coincides with that of their corresponding subsystems [25]. The analysis and design of this class of decentralized overlapping control systems has been intensively studied in the literature, primarily in the Expansion-Contraction framework [26].

In the present work, it is assumed that the given interconnected system has some distinct undesirable DFMs. A cost is assigned for establishing a link between any pair of local controllers. This can, for instance, reflect the data transmission cost required for a communication link between the control stations. The ultimate goal can be described in two steps. The first step is to characterize all the decentralized overlapping control structures with respect to which the system has no undesirable fixed modes. The second step is to determine the optimal overlapping structure which minimizes the implementation cost (associated with establishing new links between local controllers). To this end, it is first shown that the unrepeated fixed modes of the system with respect to any overlapping control structure can be identified using a graph-theoretic approach. Then, the notions of minimal sets and maximal graphs are introduced to present a simple procedure for solving the problem under study. As a by-product of the proposed development, all the decentralized overlapping control structures whose implementation cost are less than any given value and are capable of eliminating the undesirable DFMs can also be characterized efficiently.

5.2 Preliminaries

Consider a LTI interconnected system $\mathscr{S}$ consisting of ν subsystems $S_1, S_2, ..., S_\nu$ with the following state-space representation for its ith subsystem ($i \in \bar{\nu} := \{1, 2, ..., \nu\}$):

$$\begin{aligned} \dot{x}_i(t) &= A_{ii}x_i(t) + B_{ii}u_i(t) + f_i(x(t), u(t)) \\ y_i(t) &= C_{ii}x_i(t) + D_{ii}u_i(t) + g_i(x(t), u(t)) \end{aligned} \tag{5.1}$$

where $x_i(t) \in \mathfrak{R}^{n_i}$, $u_i(t) \in \mathfrak{R}^{m_i}$ and $y_i(t) \in \mathfrak{R}^{r_i}$ are the state, the input and the output of the subsystem S_i, respectively, and $f_i(x(t), u(t))$ and $g_i(x(t), u(t))$ denote the effect of the other subsystems on S_i through its incoming interconnections. Assume that:

$$\begin{aligned} f_i(x(t), u(t)) &= \sum_{j=1,\ j\neq i}^{\nu} A_{ij}x_j(t) + \sum_{j=1,\ j\neq i}^{\nu} B_{ij}u_j(t), \\ g_i(x(t), u(t)) &= \sum_{j=1,\ j\neq i}^{\nu} C_{ij}x_j(t) + \sum_{j=1,\ j\neq i}^{\nu} D_{ij}u_j(t) \end{aligned} \tag{5.2}$$

for any $i \in \bar{\nu}$. The state-space model of the system $\mathscr{S}$ can be rewritten as:

$$\begin{aligned} \dot{x}(t) &= Ax(t) + \sum_{j=1}^{\nu} B_j u_j(t) \\ y_i(t) &= C_i x(t) + \sum_{j=1}^{\nu} D_{ij}u_j(t), \quad i \in \bar{\nu} \end{aligned} \tag{5.3}$$

where A is a matrix with (i, j) block entry A_{ij}, for any $i, j \in \bar{\nu}$, and:

$$B_j = \begin{bmatrix} B_{1j}^T & B_{2j}^T & \cdots & B_{\nu j}^T \end{bmatrix}^T, \quad C_j = \begin{bmatrix} C_{j1} & C_{j2} & \cdots & C_{j\nu} \end{bmatrix}, \quad j \in \bar{\nu} \tag{5.4}$$

A structurally constrained controller for the system $\mathscr{S}$ consists of ν local controllers, partially interacting with each other. The following definition will prove convenient in formulating the interaction policy between different subsystems (or equivalently, between the local controllers).

Definition 1 *Given a structurally constrained controller, define the control interaction set* $\mathbf{K}$ *associated with this controller as a set which contains the entry* k_{ij}, $i, j \in \bar{\nu}$, *if and only if* $y_j(t)$ *can contribute to the construction of* $u_i(t)$ *in the controller.*

Define the set $\mathbf{K}_d := \{k_{11}, k_{22}, ..., k_{\nu\nu}\}$. Any controller whose structure complies with $\mathbf{K}_d$ is composed of ν isolated (non-interacting) local controllers; i.e., there is no data transmission between the local controllers. A controller with this structure will be referred to as a decentralized controller throughout the paper [2]. Moreover, any controller whose interaction set $\mathbf{K}$ is not equal to $\mathbf{K}_d$ but includes it (i.e., $\mathbf{K} \neq \mathbf{K}_d$, $\mathbf{K}_d \subset \mathbf{K}$) is called an overlapping controller [23].

The DFMs of $\mathscr{S}$ are indeed the modes of the system which are fixed with respect to all LTI controllers complying with the control interaction set $\mathbf{K}_d$ [4]. Furthermore, in the case of an overlapping controller with the interaction set $\mathbf{K}$, the DOFMs of the system $\mathscr{S}$ w.r.t. $\mathbf{K}$ are the modes of the system which are fixed under any LTI controller whose structure complies with $\mathbf{K}$ [23].

5.3 Main Results

Consider the system $\mathscr{S}$ given by (5.3), and assume that it has some distinct undesirable DFMs. It is desired to displace these undesirable fixed modes using a proper control structure, in order to meet the design specifications. Let these undesirable modes be denoted by $\sigma_1, \sigma_2, ..., \sigma_\mu$. By definition, there is no LTI controller complying with $\mathbf{K}_d$ to displace any of these modes. Hence, it is desired to expand the control interaction set $\mathbf{K}_d$ by adding another set $\mathbf{K}_e$ to it such that none of these unwanted modes will be immovable with respect to the new control interaction set $\mathbf{K}_d \cup \mathbf{K}_e$. This problem is investigated in the sequel for a particular case first, and then is extended to the general case.

5.3.1 Displacing a single unrepeated DFM

Assume that σ is an arbitrary unrepeated mode of the system $\mathscr{S}$. One possible state-space realization for this system is given by:

$$\begin{aligned} \dot{\mathbf{x}}(t) &= \begin{bmatrix} \sigma & 0 \\ 0 & \mathbf{A} \end{bmatrix} \mathbf{x}(t) + \sum_{j=1}^{\nu} \mathbf{B}_j u_j(t) \\ y_i(t) &= \mathbf{C}_i \mathbf{x}(t) + \sum_{j=1}^{\nu} \mathbf{D}_{ij} u_j(t), \quad i \in \bar{\nu} \end{aligned} \tag{5.5}$$

where the matrices $\mathbf{A}, \mathbf{B}_j, \mathbf{C}_i$ and $\mathbf{D}_{ij}$, $i, j \in \bar{\nu}$ can be obtained by using a proper similarity transformation, but their exact form is not essential in the main development (it is to be noted that DFMs are invariant under any similarity transformation). Define the matrix M as follows:

$$M := \begin{bmatrix} \mathbf{C}_1 \\ \vdots \\ \mathbf{C}_\nu \end{bmatrix} \begin{bmatrix} 0 & 0 \\ 0 & (\mathbf{A} - \sigma I)^{-1} \end{bmatrix} \begin{bmatrix} \mathbf{B}_1 & \cdots & \mathbf{B}_\nu \end{bmatrix} - \begin{bmatrix} \mathbf{D}_{11} & \cdots & \mathbf{D}_{1\nu} \\ \vdots & \ddots & \vdots \\ \mathbf{D}_{\nu 1} & \cdots & \mathbf{D}_{\nu\nu} \end{bmatrix} \tag{5.6}$$

Note that since the multiplicity of σ is assumed to be 1, it is not an eigenvalue of the matrix $\mathbf{A}$. Denote the (i,j) block entry of M with $M_{ij} \in \Re^{r_i \times m_j}$, for any $i,j \in \bar{\nu}$.

Procedure 1 *[18] Construct a bipartite graph $\mathscr{G}$ with two sets of vertices $\mathscr{V}$ (set 1) and $\bar{\mathscr{V}}$ (set 2) and the tagged vertices $1,2,\ldots,\nu$ in each of the two sets. For any $i,j \in \bar{\nu}$, carry out the following steps:*

1) Connect vertex j of the set $\mathscr{V}$ to vertex i of the set $\bar{\mathscr{V}}$ if $M_{ij} = 0$.
2) Mark vertex i of the set $\mathscr{V}$ if the first column of the matrix $\mathbf{C}_i$ is a zero vector.
3) Mark vertex j of the set $\bar{\mathscr{V}}$ if the first row of the matrix $\mathbf{B}_j$ is a zero vector.

Definition 2 *Consider an arbitrary graph $\hat{\mathscr{G}}$ with ζ vertices in any of its two sets, labeled $1,2,...,\zeta$. A subgraph of $\hat{\mathscr{G}}$ is said to span the vertices of $\hat{\mathscr{G}}$, if the labels of its vertices are distinct and form the set $\{1,2,...,\zeta\}$.*

Identify every subgraph of $\mathscr{G}$ which satisfies the following criteria:

i) It is a complete bipartite subgraph.
ii) All of its vertices are marked.
iii)It spans the vertices of the graph $\mathscr{G}$.

Denote all such subgraphs with $\mathscr{G}_1, \mathscr{G}_2, ..., \mathscr{G}_w$. Moreover, denote set 1 and set 2 (see Procedure 1) of the graph $\mathscr{G}_j$ with $\mathscr{V}_j$ and $\bar{\mathscr{V}}_j$, respectively, for any $j \in \{1,2,...,w\}$.

The following lemma is elicited from [18].

Lemma 1 *The mode σ is a DFM of the system $\mathscr{S}$ if and only if the nonnegative integer w is strictly positive.*

Assume for now that w is strictly positive, and consequently the mode σ is fixed with respect to any LTI controller complying with $\mathbf{K}_d$. It is desired to obtain all overlapping control structures which are able to displace this mode.

Procedure 2 *For any given set $\{k_{i_1 j_1}, k_{i_2 j_2}, ..., k_{i_z j_z}\}$, form a bipartite graph $\mathscr{G}(\{k_{i_1 j_1}, k_{i_2 j_2}, ..., k_{i_z j_z}\})$ as follows:*

- *Put $\nu + z$ vertices in set 1 and set 2 of the graph $\mathscr{G}(\{k_{i_1 j_1}, ..., k_{i_z j_z}\})$.*
- *Assign the labels $1,2,...,\nu, j_1, j_2, ..., j_z$ to the vertices of set 1.*
- *Assign the labels $1,2,...,\nu, i_1, i_2, ..., i_z$ to the vertices of set 2.*
- *Consider any two arbitrary vertices of the graph $\mathscr{G}(\{k_{i_1 j_1}, ..., k_{i_z j_z}\})$ which do not pertain to the same set of vertices. Let the labels of these two vertices be λ_1 (in set 1) and λ_2 (in set 2). Connect these two vertices to each other in the graph $\mathscr{G}(\{k_{i_1 j_1}, ..., k_{i_z j_z}\})$ if and only if there is an edge between vertex λ_1 of $\mathscr{V}$ and vertex λ_2 of $\bar{\mathscr{V}}$ in the graph $\mathscr{G}$.*

It is notable that some labels in the graph $\mathscr{G}(\{k_{i_1 j_1}, k_{i_2 j_2}, ..., k_{i_z j_z}\})$ are recurrent. The next theorem proposes a simple method to verify whether or not the mode σ is a DOFM of the system $\mathscr{S}$ with respect to a given control interaction set.

Theorem 1 *Given the set* $\{k_{p_1q_1}, k_{p_2q_2}, ..., k_{p_\alpha q_\alpha}\}$*, the mode* σ *is not a DOFM of the system* $\mathscr{S}$ *w.r.t.* $\mathbf{K}_d \cup \{k_{p_1q_1}, ..., k_{p_\alpha q_\alpha}\}$ *if and only if the graph* $\mathscr{G}(\{k_{p_1q_1}, ..., k_{p_\alpha q_\alpha}\})$ *does not contain a complete bipartite subgraph with all vertices marked, which spans its vertices (see Definition 2).*

Sketch of the proof: The proof will be given here for $\alpha = 1$, as its generalization is straightforward. By obtaining two transformation matrices discussed in [23] and pursuing the approach given therein, it can be easily verified that σ is a DOFM of the system $\mathscr{S}$ w.r.t. the control interaction set $\mathbf{K}_d \cup \{k_{p_1q_1}\}$ if and only if it is a DFM of the following system:

$$
\begin{aligned}
\dot{\tilde{x}}(t) &= A\tilde{x}(t) + \sum_{j=1}^{v} B_j\tilde{u}_j(t) + B_{p_1}\tilde{u}_{v+1}(t) \\
\tilde{y}_i(t) &= C_i\tilde{x}(t) + \sum_{j=1}^{v} D_{ij}\tilde{u}_j(t) + D_{ip_1}\tilde{u}_{v+1}(t), \quad i \in \bar{v} \\
\tilde{y}_{v+1}(t) &= \tilde{C}_{q_1}\tilde{x}(t) + \sum_{j=1}^{v} D_{q_1j}\tilde{u}_j(t) + D_{q_1p_1}\tilde{u}_{v+1}(t)
\end{aligned}
\qquad (5.7)
$$

Note that this system has one input and one output more than the system $\mathscr{S}$. The proof follows by applying the graph-theoretic approach given in [18] (which was explained in Lemma 1 for the system $\mathscr{S}$) to the system given in (5.7). ■

So far, it is shown how the existence of a DOFM can be concluded from a bipartite graph. This result will be used next to characterize all desirable overlapping control structures.

Definition 3 *The set* $\{k_{p_1q_1}, k_{p_2q_2}, ..., k_{p_\alpha q_\alpha}\}$ *is said to be minimal w.r.t.* σ *if and only if the mode* σ *is not a DOFM of the system* $\mathscr{S}$ *w.r.t.* $\mathbf{K}_d \cup \{k_{p_1q_1}, k_{p_2q_2}, ..., k_{p_\alpha q_\alpha}\}$*, while it is a DOFM of* $\mathscr{S}$ *w.r.t.* $\mathbf{K}_d \cup \{k_{p_1q_1}, ..., k_{p_{j-1}q_{j-1}}, k_{p_{j+1}q_{j+1}}, ..., k_{p_\alpha q_\alpha}\}$ *for any* $j \in \{1, 2, ..., \alpha\}$.

Definition 4 *A subgraph of the graph* $\mathscr{G}$ *is said to be maximal if:*

i) It is a complete bipartite subgraph.
ii) All of its vertices are marked.
iii) The set of the labels of its vertices is equal to the set $\bar{v}$ *(note that this condition is slightly different from spanning the vertices, as the labels can be recurrent here).*
iv) The graph $\mathscr{G}$ *has no other subgraph satisfying criteria (i), (ii) and (iii) given above such that it includes this subgraph.*

Using the combinatorial algorithms, the maximal subgraphs of $\mathscr{G}$ can be easily identified (analogously to the algorithms for finding the complete bipartite graphs with maximum number of edges). Denote such subgraphs with $\tilde{\mathscr{G}}_1, \tilde{\mathscr{G}}_2, ..., \tilde{\mathscr{G}}_{\tilde{w}}$. Moreover, denote set 1 and set 2 of vertices of the graph $\tilde{\mathscr{G}}_j$ with $\tilde{\mathscr{V}}_j$ and $\bar{\tilde{\mathscr{V}}}_j$, respectively, for any $j \in \{1, 2, ..., \tilde{w}\}$. It is to be noted that the number $\tilde{w}$ is typically small, due to the generic property of the fixed modes. The subsequent remark aims to present a bound on the number $\tilde{w}$.

Remark 1 *From the definition of a maximal graph, the sets of vertices $\tilde{\mathscr{V}}_1, ..., \tilde{\mathscr{V}}_{\tilde{w}}$ are all distinct. Moreover, it is straightforward to show that one of the sets $\mathscr{V}_1, \mathscr{V}_2, ..., \mathscr{V}_w$ is exactly the same as $\tilde{\mathscr{V}}_j$, for any $j \in \{1,2,...,\tilde{w}\}$. These two facts point to the inequality $\tilde{w} \leq w$.*

Theorem 2 *Assume that the set $\{k_{p_1q_1}, k_{p_2q_2}, ..., k_{p_\alpha q_\alpha}\}$ is minimal w.r.t. the mode σ. Then, the number α is less than or equal to $\tilde{w}$.*

Proof: From the definition of a minimal set, the mode σ is a DOFM of the system $\mathscr{S}$ w.r.t. the control interaction set $\mathbf{K}_d \cup \{k_{p_1q_1}, ..., k_{p_{j-1}q_{j-1}}, k_{p_{j+1}q_{j+1}}, ..., k_{p_\alpha q_\alpha}\}$, for any $j \in \{1,2,...,\alpha\}$. Hence, it can be concluded from Theorem 1 that the graph $\mathscr{G}(\{k_{p_1q_1}, ..., k_{p_{j-1}q_{j-1}}, k_{p_{j+1}q_{j+1}}, ..., k_{p_\alpha q_\alpha}\})$ has a complete bipartite subgraph with marked vertices, which spans the vertices of the graph. This subgraph should include either the duplicated vertex q_i in its set 1 or the duplicated vertex p_i in its set 2, for all $i \in \{1,2,...,j-1,j+1,...,\alpha\}$. On the other hand, it is straightforward to show that there exists an integer $f_j \in \{1,...,\tilde{w}\}$ such that this subgraph is included in $\tilde{\mathscr{G}}_{f_j}$ (in light of the definition of a maximal graph). Thus, one comes to the conclusion immediately that the following logic statement is true:

$$\left(q_i \in \tilde{\mathscr{V}}_{(f_j)}\right) \vee \left(p_i \in \bar{\tilde{\mathscr{V}}}_{(f_j)}\right), \ \forall i \in \{1,2,...,j-1,j+1,...,\alpha\} \tag{5.8}$$

where $\vee$ is the logic operation *OR*. Now, to prove Theorem 2 by contradiction, assume that $\tilde{w} < \alpha$. Since all the natural numbers $f_1, f_2, ..., f_\alpha$ belong to the set $\{1,2,...,\tilde{w}\}$ and also the inequality $\tilde{w} < \alpha$ holds, it can be concluded from the Dirichlet's Principle that at least two of the values $f_1, f_2, ..., f_\alpha$ are identical. Without any loss of generality, assume that $f_1 = f_2 = f$ for some positive number f. Consider the relation (5.8) for the values $j = 1$ and $j = 2$. The amalgamation of these two sets of relations will arrive at the following true statement:

$$\left(q_i \in \tilde{\mathscr{V}}_f\right) \vee \left(p_i \in \bar{\tilde{\mathscr{V}}}_f\right), \ \forall i \in \{1,2,...,\alpha\} \tag{5.9}$$

The relation (5.9) yields that the graph $\mathscr{G}(\{k_{p_1q_1}, k_{p_2q_2}, ..., k_{p_\alpha q_\alpha}\})$ includes a complete bipartite subgraph with the properties pointed out in Theorem 1. This implies that the mode σ is a DOFM w.r.t. $\mathbf{K}_d \cup \{k_{p_1q_1}, k_{p_2q_2}, ..., k_{p_\alpha q_\alpha}\}$, which contradicts the original assumption of minimality. ∎

Theorem 2 states that if by adding more than $\tilde{w}$ communication links to the decentralized control structure the mode σ is no longer fixed, then some of the links are redundant and have no contribution in displacing the mode. It is worth mentioning that the result of Theorem 2 significantly diminishes the computational burden of finding all the minimal sets. One can use the following theorem to develop an algorithm for finding the minimal sets systematically.

Theorem 3 *The set $\{k_{p_1q_1}, k_{p_2q_2}, ..., k_{p_\alpha q_\alpha}\}$ is minimal w.r.t. σ if and only if the criteria given below both hold:*

- *For any $j \in \{1,2,...,\alpha\}$, there exists an integer $f_j \in \{1,...,\tilde{w}\}$ such that the statements:*

$$\begin{aligned} &\left(q_i \in \tilde{\mathscr{V}}_{(f_j)}\right) \vee \left(p_i \in \bar{\tilde{\mathscr{V}}}_{(f_j)}\right), \ \forall i \in \{1,2,...,j-1,j+1,...,\alpha\} \\ &\left(q_j \notin \tilde{\mathscr{V}}_{(f_j)}\right) \wedge \left(p_j \notin \bar{\tilde{\mathscr{V}}}_{(f_j)}\right) \end{aligned} \tag{5.10}$$

are true, where $f_1, f_2, ..., f_{\tilde{w}}$ are all distinct (note that $\wedge$ is the logic operation AND).

- *There exists no integer $f \in \{1,...,\tilde{w}\}$ such that the following logic statement is true:*

$$\left(q_i \in \tilde{\mathscr{V}}_f\right) \vee \left(p_i \in \bar{\tilde{\mathscr{V}}}_f\right), \ \forall i \in \{1,2,...,\alpha\} \tag{5.11}$$

Proof: The proof of this theorem follows directly from the discussions given in the proof of Theorem 2. The details are omitted here. ■

Theorem 3 implicitly proposes a simple method to compute all the minimal sets w.r.t. to the fixed mode σ.

5.3.2 Displacing multiple unrepeated DFMs

The methodology presented in the preceding subsection will be deployed here to characterize all the control interaction sets $\mathbf{K}_e$ such that the DFMs $\sigma_1, \sigma_2, ..., \sigma_\mu$ are all movable w.r.t to $\mathbf{K}_d \cup \mathbf{K}_e$. Although the mode σ_i, $i \in \{1,2,...,\mu\}$, is by assumption an unrepeated DFM of the system $\mathscr{S}$, its multiplicity as a regular mode of the system can be greater than 1. In this case, the aforementioned method cannot be applied to the relevant problem directly. To obviate this issue, one can consider a generic static decentralized controller and apply it to the system $\mathscr{S}$ so that the multiplicity of the mode σ_i, $i \in \{1,2,...,\mu\}$, will be exactly equal to 1 in the resultant system [4]. Therefore, with no loss of generality, assume henceforth that the mode σ_i, $i \in \{1,2,...,\mu\}$, is not only an unrepeated DFM but also an unrepeated mode of the system $\mathscr{S}$.

For any $i \in \{1,2,...,\mu\}$, obtain all minimal sets associated with the mode σ_i using the approach given in the previous subsection, and denote them with $\mathbf{K}_e^{i,1}, \mathbf{K}_e^{i,2}, ..., \mathbf{K}_e^{i,z_i}$. The following corollary states how the underlying problem can be treated.

Corollary 1 *Given the control interaction set $\mathbf{K}_e$, none of the modes $\sigma_1, \sigma_2, ..., \sigma_\mu$ are DOFMs of the system $\mathscr{S}$ w.r.t. $\mathbf{K}_d \cup \mathbf{K}_e$ if and only if there exist integers $g_1, g_2, ..., g_\mu$ with the following property:*

$$\{\mathbf{K}_e^{1,g_1} \cup \mathbf{K}_e^{2,g_2} \cup \cdots \cup \mathbf{K}_e^{\mu,g_\mu}\} \subseteq \mathbf{K}_e \tag{5.12}$$

Proof: The proof follows immediately from the definition of a minimal set. ■

In practice, it is desired that the addition of the set of interconnection links $\mathbf{K}_e$ to the control structure be as inexpensive as possible. In order to take the expenditure of this inclusion into account, it is assumed that the cost of implementing the communication link k_{ij} is prespecified by the designer, and is denoted by $\mathscr{C}_{ij}$, for any $i, j \in \bar{\nu}$. Note that this cost should normally be defined in terms of the exterior factors associated with the system such as the distance between the subsystems or the nature of the outputs to be transmitted to the other subsystems. This restriction results from the fact that different state-space representations are used throughout the paper for different modes, which may lead to the inconsistency in the cost evaluation. By virtue of Corollary 1, finding the least costly $\mathbf{K}_e$ with the aforementioned property (displacing certain fixed modes) can be translated into obtaining all the sets $\mathbf{K}_e$ representable as $\mathbf{K}_e^{1,g_1} \cup \mathbf{K}_e^{2,g_2} \cup \cdots \cup \mathbf{K}_e^{\mu,g_\mu}$ for some integers $g_1, g_2, ..., g_\mu$, and computing their associated costs accordingly to determine which one is the least expensive.

5.4 Numerical example

Let $\mathscr{S}$ be a system consisting of four single-input single-output (SISO) subsystems with the following decoupled state-space matrices:

$$A = \begin{bmatrix} 1 & 0 & 0 & 0 \\ 0 & 2 & 0 & 0 \\ 0 & 0 & 3 & 0 \\ 0 & 0 & 0 & 4 \end{bmatrix}, \; B_1 = \begin{bmatrix} 3 \\ 4 \\ 0 \\ 1 \end{bmatrix}, \; B_2 = \begin{bmatrix} 0 \\ 2 \\ 0 \\ 6 \end{bmatrix}, \; B_3 = \begin{bmatrix} 0 \\ 7 \\ 9 \\ -5 \end{bmatrix}, \; B_4 = \begin{bmatrix} 0 \\ 0 \\ 8 \\ 7 \end{bmatrix} \tag{5.13}$$

and

$$\begin{aligned} &C_1 = \begin{bmatrix} 0 & 2 & 4 & 3 \end{bmatrix}, \; C_2 = \begin{bmatrix} 0 & -6 & 0 & 8 \end{bmatrix}, \; C_3 = \begin{bmatrix} 0 & 4 & 0 & -9 \end{bmatrix}, \\ &C_4 = \begin{bmatrix} 5 & 1 & 0 & 7 \end{bmatrix}, \; D_{11} = -5, \; D_{12} = 10, \; D_{13} = 27, \; D_{14} = 23, \; D_{21} = 32, \\ &D_{22} = 60, \; D_{23} = -3, \; D_{24} = 56/3, \; D_{31} = -25, \; D_{32} = -62, \; D_{33} = 43, \\ &D_{34} = -21, \; D_{41} = -4.5, \; D_{42} = 40, \; D_{43} = 16, \; D_{44} = 7, \end{aligned} \tag{5.14}$$

Consider now the mode $\sigma = 1$. The graph $\mathscr{G}$ associated with this mode is depicted in Figure 5.1(a). This graph contains a complete bipartite subgraph with vertex 1 from set 1 of $\mathscr{G}$ and vertices 2, 3 and 4 from set 2 of $\mathscr{G}$ such that its vertices are all marked and also it spans the vertices of $\mathscr{G}$. Therefore, it results from Lemma 1 that the mode 1 is a DFM of the system. Likewise, it can be shown that the mode 3 is also a DFM of the system, while the modes 2 and 4 are not. Note that the control interaction set corresponding to a decentralized controller in this example is equal to $\mathbf{K}_d = \{k_{11}, k_{22}, k_{33}, k_{44}\}$.

It is desired now to expand the structure of the control system from decentralized to overlapping so that both of these unstable DFMs can be displaced. Let the mode $\sigma = 1$ be treated first. It is straightforward to obtain the relation $w = 2$ from the graph $\mathscr{G}$ in Figure 5.1(a). Therefore, the subgraphs $\mathscr{G}_1$ and $\mathscr{G}_2$ will have the following sets of vertices:

$$\mathscr{V}_1 = \{1\}, \quad \bar{\mathscr{V}}_1 = \{2,3,4\}, \quad \mathscr{V}_2 = \{1,2,3\}, \quad \bar{\mathscr{V}}_2 = \{4\} \tag{5.15}$$

On the other hand, the graph $\mathscr{G}$ has two maximal subgraphs which are the same as $\mathscr{G}_1$ and $\mathscr{G}_2$. It can now be concluded from Theorem 3 that the minimal sets w.r.t. $\sigma = 1$ are:

$$\mathbf{K}_e^{1,1} = \{k_{14}\}, \quad \mathbf{K}_e^{1,2} = \{k_{12}, k_{34}\}, \quad \mathbf{K}_e^{1,3} = \{k_{13}, k_{24}\} \tag{5.16}$$

Note that as expected from Theorem 2, these sets have at most 2 elements, due to the relation $\tilde{w} = 2$. Analogously, the minimal sets w.r.t. $\sigma = 3$ can be obtained as:

$$\mathbf{K}_e^{3,1} = \{k_{31}\}, \quad \mathbf{K}_e^{3,2} = \{k_{41}\} \tag{5.17}$$

(note that $\tilde{w}$ for the mode $\sigma = 3$ is equal to 1). It results from Corollary 1 that the modes 1 and 3 are not DOFMs of the system w.r.t. the control interaction $\mathbf{K}_d \cup \mathbf{K}_e$ if and only if the following condition is satisfied for the set $\mathbf{K}_e$:

$$\exists \zeta_1 \in \{1,2,3\}, \exists \zeta_2 \in \{1,2\}: \quad \{\mathbf{K}_e^{1,\zeta_1} \cup \mathbf{K}_e^{3,\zeta_2}\} \subseteq \mathbf{K}_e \tag{5.18}$$

Assume now that all of the communication links have the same cost, i.e., $\mathscr{C}_{ij} = 1$, $i, j \in \{1,2,3,4\}$. In this case, the least costly $\mathbf{K}_e$ will be $\{k_{14}, k_{31}\}$ or $\{k_{14}, k_{41}\}$ with the implementation cost of 2. The graph $\mathscr{G}(\{k_{14}, k_{31}\})$ corresponding to the modes 1 and 3 are depicted in Figures 5.1(b) and 5.1(c), respectively. It can be easily verified that none of these graphs has a subgraph with the properties mentioned in Theorem 1. This validates the obtained result stating that the modes 1 and 3 are not the DOFMs of the system $\mathscr{S}$ w.r.t. $\mathbf{K}_d \cup \{k_{14}, k_{31}\}$.

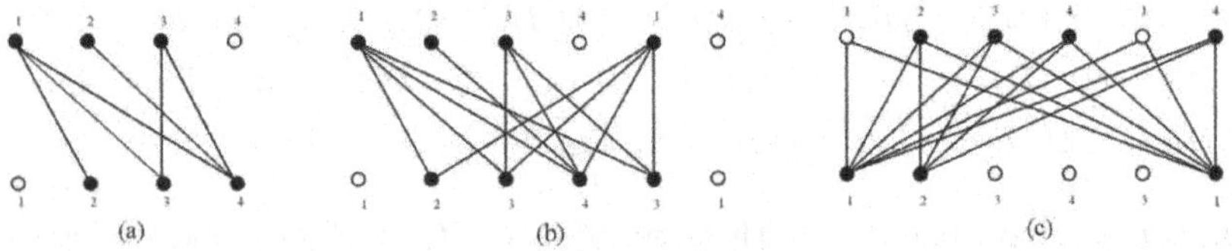

Fig. 5.1 a) The graph $\mathscr{G}$ associated with the mode $\sigma = 1$; b) the graph $\mathscr{G}(\{k_{14}, k_{31}\})$ corresponding to the mode $\sigma = 1$; c) the graph $\mathscr{G}(\{k_{14}, k_{31}\})$ corresponding to the mode $\sigma = 3$.

To show the effect of implementation expenditure on obtaining the set $\mathbf{K}_e$, assume that:

$$\mathscr{C}_{14} = 5, \ \mathscr{C}_{12} = 2, \ \mathscr{C}_{34} = 2, \ \mathscr{C}_{13} = 1, \ \mathscr{C}_{24} = 1, \ \mathscr{C}_{31} = 5, \ \mathscr{C}_{41} = 4 \tag{5.19}$$

In this case, the set $\mathbf{K}_e$ will be equal to $\{k_{13}, k_{24}, k_{41}\}$ with the implementation cost of 6.

5.5 Summary

This chapter tackles the stabilizability problem for an interconnected system with a number of distinct undesirable DFMs, by means of the structurally constrained controllers. It is well-known that a LTI decentralized controller comprising a set of isolated local controllers cannot displace any DFMs. Thus, the objective of this chapter is to establish some interactions between the local controllers in order to displace the undesirable DFMs. To this end, the knowledge of the system is transformed into a number of bipartite graphs (corresponding to the unwanted DFMs). Subsequently, the notions of minimal sets of interactions and maximal subgraphs are introduced. A simple procedure is then proposed to characterize all the possible sets of interactions which maintain the mentioned property. The numerical example provided elucidates the efficacy of the present work.

References

1. J. Lavaei, A. Momeni and A. G. Aghdam, "High-performance decentralized control for formation flying with leader-follower structure," in *Proceedings of 45th IEEE Conference on Decision and Control*, San Diego, CA, 2006.
2. J. Lavaei and A. G. Aghdam, "Decentralized control design for interconnected systems based on a centralized reference controller," in *Proceedings of 45th IEEE Conference on Decision and Control*, San Diego, CA, 2006.
3. A. B. Ozguler, "Global stabilization via local stabilizing actions," *IEEE Transactions on Automatic Control*, vol. 51, no. 3, pp. 530-533, 2006.
4. E. J. Davison and T. N. Chang, "Decentralized stabilization and pole assignment for general proper systems," *IEEE Transactions on Automatic Control*, vol. 35, no. 6, pp. 652-664, 1990.
5. D. D. Šiljak, Decentralized control of complex systems, Cambridge: Academic Press, 1991.
6. H. Li and Y. Wang, "Nonlinear robust decentralized control of multimachine power systems," *in Proc. 7th International Power Engineering Conference*, pp. 1-6, 2005.
7. M. E. Khatir and E. J. Davison, "Decentralized control of a large platoon of vehicles operating on a plane with steering dynamics," *in Proc. 2005 American Control Conference*, Portland, OR, pp. 2159-2165, 2005.
8. A. Fay and I. Fischer, "Decentralized control strategies for transportation systems," *International Conference on Control and Automation*, Budapest, Hungary, pp. 898-903, 2005.
9. B. A. Movsichoff, C. M. Lagoa, and H. Che, "Decentralized optimal traffic engineering in connectionless networks," *IEEE Journal on Selected Areas in Communications*, vol. 23, no. 2, pp. 293-303, 2005.
10. J. Lavaei and A. G. Aghdam, "Elimination of fixed modes by means of high-performance constrained periodic control," in *Proceedings of 45th IEEE Conference on Decision and Control*, San Diego, CA, 2006.
11. J. Lavaei and A. G. Aghdam, "Simultaneous LQ control of a set of LTI systems using constrained generalized sampled-data hold functions," *Automatica*, vol. 43, no. 2, pp. 274-280, 2007.
12. Z. Duan, J. Z. Wang, and L. Huang, "Special decentralized control problems and effectiveness of parameter-dependent Lyapunov function method," *in Proc. 2005 American Control Conference*, vol. 3, pp. 1697-1702, 2005.
13. J. Lavaei and A. G. Aghdam, "Optimal periodic feedback design for continuous-time LTI systems with constrained control structure," *International Journal of Control*, vol. 80, no. 2, pp. 220-230, 2007.

14. S. H. Wang and E. J. Davison, "On the stabilization of decentralized control systems," *IEEE Transactions on Automatic Control*, vol. 18, no. 5, pp. 473-478, 1973.
15. B. D. O. Anderson and D. J. Clements, "Algebraic characterizations of fixed modes in decentralized systems," *Automatica*, vol. 17, no. 5, pp. 703-712, 1981.
16. B. D. O. Anderson, "Transfer function matrix description of decentralized fixed modes," *IEEE Transactions on Automatic Control*, vol. 27, no. 6, pp. 1176-1182, 1982.
17. E. J. Davison and S. H. Wang, "A characterization of decentralized fixed modes in terms of transmission zeros," *IEEE Transactions on Automatic Control*, vol. 30, no. 1, pp. 81-82, 1985.
18. J. Lavaei and A. G. Aghdam, "Characterization of decentralized and quotient fixed modes via graph theory," in *Proceedings of 2007 American Control Conference*, New York, NY, 2007.
19. V. Armentano and M. Singh, "A procedure to eliminate decentralized fixed modes with reduced information exchange," *IEEE Transactions on Automatic Control*, vol. 27, no. 1, pp. 258-260, 1982.
20. K. Unyelioglu and E. Sezer, "Optimum feedback patterns in multivariable control systems," *International Journal of Control*, vol. 49, no. 3, pp. 791-808, 1989.
21. M. E. Sezer and D. D. Šiljak, "Structurally fixed modes," *Systems & Control Letters*, vol. 1, no. 1, pp. 60-64, 1981.
22. A. Belmehdi and D. Boukhetala, "Method to eliminate structurally fixed modes in decentralized control systems," *International Journal of Systems Science*, vol. 33, no. 15, pp. 1249-1256, 2002.
23. J. Lavaei and A. G. Aghdam, "A necessary and sufficient condition for the existence of a LTI stabilizing decentralized overlapping controller," in *Proceedings of 45th IEEE Conference on Decision and Control*, San Diego, CA, 2006.
24. J. Lavaei and A. G. Aghdam, "On structurally constrained control design with a prespecified form," in *Proceedings of 2007 American Control Conference*, New York, NY, 2007.
25. D. D. Šiljak and A. I. Zecevic, "Control of large-scale systems: Beyond decentralized feedback," *Annual Reviews in Control*, vol. 29, no.2, pp. 169-179, 2005.
26. M. Ikeda, D. D. Šiljak and D. E. White, "Decentralized control with overlapping information sets," *Journal of Optimization Theory and Applications*, vol. 34, no. 2, pp. 279-310, 1981.

Chapter 6
Characterization of Stabilizing Structurally Constrained Controllers

6.1 Abstract

The focus of this chapter is directed towards the problem of characterizing the information flow structures of all classes of LTI structurally constrained controllers with respect to which a given interconnected system has no fixed modes. Any class of structurally constrained controllers can be described by a set of communication links, which delineates how the local controllers of any controller in that class interact with each other. To achieve the objective, a cost is first attributed for establishing any communication link in the control structure. These costs are part of deign specifications and represent the expenditure of data transmission between different subsystems. A simple graph-theoretic method is then proposed to characterize all the relevant classes of controllers systematically. As a by-product of this approach, all classes of LTI stabilizing structurally constrained controllers with the minimum implementation cost are attained using a novel algorithm. The primary advantages of this approach are its simplicity and computational efficiency. The efficacy and importance of this work are thoroughly illustrated in a numerical example. This chapter integrates the ideas proposed in a recently published work and some original techniques to develop its main results.

6.2 Introduction

A great number of real-world plants can be regarded as interconnected systems with several interacting subsystems [1]. A typical controller for an interconnected system is composed of a set of local controllers corresponding to different subsystems. Normally, each local controller should receive some information from all the subsystems in order for the resultant controller to achieve best possible performance. This case is referred to as the centralized control strategy in the literature. Most of the control design techniques spontaneously arrive at centralized controllers, which may have practical problems as far as control implementation is concerned [2]. More precisely, there are some practical issues which may hinder employing a general centralized controller for an interconnected system. The primary reason is that the transmission of information from a subsystem to a local controller of another subsystem may be infeasible or quite costly. Such problems appear, for instance, in formation flying where the shadow phenomenon occurs for a specific time interval [3]. The case when some transmission links are remarkably costly comes about in the systems whose subsystems are geographically remote, e.g., in a power system with several stations in different cities. Furthermore, for a system consisting of several subsystems, the computational complexity associated with the centralized control structure can be quite high. These practical restrictions introduce the motivation for utilizing structurally constrained controllers [4].

Decentralized control is a particular type of structurally constrained controllers which has attracted a considerable amount of interest in the control community. A decentralized controller is the union of a number of local controllers which do not exchange information [5, 6]. Decentralized control theory has found applications in a wide range of real-world systems such as communication networks, power systems and traffic networks. Different decentralized design techniques have thoroughly been investigated and well-documented [7, 8, 9, 10].

In many control applications, there may exist overlapping between certain subsystems of an interconnected system. In such systems, it is often desired to design a structurally constrained controller whose local controllers partially interact with each other with the same overlapping topology as their corresponding subsystems. This conceptual notion is envisaged as decentralized overlapping control strategy in the literature. This class of structurally constrained controllers has been studied intensively, mostly in the framework of Expansion-Inclusion principle [11, 12, 13].

The most important problem in conjunction with the structurally constrained control design is the stabilizability verification. To address this problem, the notion of a decentralized fixed mode (DFM) was introduced in [14] to identify the modes of the system which are fixed with respect to any LTI decentralized controller. Several methods are proposed accordingly to characterize the DFMs of a system efficiently. For instance, the paper [7] proposes a simple graph-theoretic approach to obtain the unrepeated DFMs of a system without having to experience numerical difficulties. Since a DFM may be eliminated by means of a nonlinear or time-varying decentralized controller, the notion of a quotient fixed mode (QFM) was introduced in [15] to characterize all the modes of a system which are immovable with respect to any nonlinear and time-varying decentralized controller.

More recently, the notion of a decentralized overlapping fixed mode (DOFM) was introduced in [16] to mathematically describe the fixed modes of a system with respect to any given class of structurally constrained LTI controllers. A method is also proposed in [16] to obtain the DOFMs of the system efficiently. One should take note of the fact that DOFMs do not necessarily correspond to the overlapping systems discussed earlier, and is defined for any arbitrary system. In addition, the notion of a quotient decentralized overlapping fixed mode (QOFM) was introduced in [17] as an extension of the notion of QFMs to the overlapping control structure.

Consider an interconnected system with ν subsystems. It can be easily verified that there exist $2^{(\nu^2)} - 1$ classes of structurally constrained controllers for this system. Since this number grows exponentially with ν, choosing the classes which fit into the control objectives may be a formidable task. It is worth mentioning that several approaches have been proposed in the literature for the design of structurally constrained controllers to achieve any objective such as pole-placement or LQ optimality. However, all of these methods require that the structure of the desired controller be known *a priori*.

Given an interconnected system, the focal problem of this chapter is to find all classes of LTI structurally constrained controllers with respect to which the system has no fixed modes. To handle this problem, the main concept of the graph-theoretic approach of the recent paper [18] has been exploited. Note that the work [18] addresses the problem of eliminating undesirable DFMs of a system, by adding some transmission links between the local controllers. Modified definitions of the notions of *maximal graph* and *minimal set* introduced in [18] are utilized here to address the underlying problem. Moreover, a cost is attributed to establish a communication link between any pair of local controllers. The classes of LTI stabilizing structurally constrained controllers with the minimum implementation cost are then obtained. This is achieved by proposing an efficient method which avoids unnecessary computations. The technique introduced in this chapter is also applicable to the approach given in [18] for finding the optimal control interaction sets which eliminate the unwanted DFMs.

6.3 Problem formulation

Consider a LTI interconnected system $\mathscr{S}$ consisting of ν subsystems $S_1, S_2, ..., S_\nu$. Let the ith subsystem of $\mathscr{S}$, $i \in \bar{\nu} := \{1, 2, ..., \nu\}$, be modeled as:

$$\begin{aligned} \dot{x}_i(t) &= A_{ii}x_i(t) + B_{ii}u_i(t) + f_i(x(t), u(t)) \\ y_i(t) &= C_{ii}x_i(t) + D_{ii}u_i(t) + g_i(x(t), u(t)) \end{aligned} \tag{6.1}$$

where $x_i(t) \in \Re^{n_i}$, $u_i(t) \in \Re^{m_i}$ and $y_i(t) \in \Re^{r_i}$ are the state, the input and the output of the subsystem S_i, respectively. Moreover, $f_i(x(t), u(t))$ and $g_i(x(t), u(t))$ in the above state-space representation are the interconnection signals which account for the effect of different subsystems on S_i through its incoming interconnections. Assume that these interconnection signals can be represented by:

$$\begin{aligned} f_i(x(t), u(t)) &= \sum_{j=1,\ j \neq i}^{\nu} A_{ij}x_j(t) + \sum_{j=1,\ j \neq i}^{\nu} B_{ij}u_j(t), \\ g_i(x(t), u(t)) &= \sum_{j=1,\ j \neq i}^{\nu} C_{ij}x_j(t) + \sum_{j=1,\ j \neq i}^{\nu} D_{ij}u_j(t) \end{aligned} \tag{6.2}$$

for any $i \in \bar{\nu}$. Now, the model of the system $\mathscr{S}$ will be as follows:

$$\begin{aligned} \dot{x}(t) &= Ax(t) + \sum_{j=1}^{\nu} B_j u_j(t) \\ y_i(t) &= C_i x(t) + \sum_{j=1}^{\nu} D_{ij} u_j(t), \quad i \in \bar{\nu} \end{aligned} \tag{6.3}$$

where:

$$\begin{aligned} B_j &= \left[B_{1j}^T \; B_{2j}^T \; \cdots \; B_{\nu j}^T \right]^T, \\ C_j &= \left[C_{j1} \; C_{j2} \; \cdots \; C_{j\nu} \right], \quad j \in \bar{\nu} \end{aligned} \tag{6.4}$$

Denote the modes of the system $\mathscr{S}$ with $\sigma_1, \sigma_2, ..., \sigma_n$, and assume that all the modes are distinct. As a consequence of this assumption, one can suppose with no loss of generality that the matrix A is equal to:

$$A = \text{diag}([\sigma_1 \; , \; \sigma_2 \; , \; ... \; , \; \sigma_n]) \tag{6.5}$$

(this can be achieved by using a proper similarity transformation, if necessary). Any structurally constrained controller for the system $\mathscr{S}$ comprises ν control agents corresponding to various subsystems as well as a number of transmission links. Each transmission link provides the output of a certain subsystem to a specific control agent which will be used to construct its control signal. The symbol k_{ij} $(i, j \in \bar{\nu})$ is used throughout this chapter to describe the link which transmits the output of the jth subsystem to the ith control agent (or equivalently, the ith subsystem).

Any class of structurally constrained controllers is formulated in [18] via a set, referred to as *control interaction set*. This concept will be clarified in the following definition.

Definition 1 *Given a class of structurally constrained controllers, define its associated control interaction set* $\mathbf{K}$ *as a set which includes only the symbols* k_{ij}, $i, j \in \bar{\nu}$ *whose corresponding transmission links exist in the control structure.*

Consider any arbitrary control interaction set $\mathbf{K}$. It is well-known that certain modes of the system $\mathscr{S}$ may be fixed under all LTI controllers belonging to the class of structurally constrained controllers defined by $\mathbf{K}$. These fixed modes (if any) are referred to as decentralized overlapping fixed modes (DOFM) w.r.t. (with respect to) $\mathbf{K}$.

Assume henceforth that the system $\mathscr{S}$ is controllable and observable (the results obtained can be easily extended to the case when the system is detectable and stabilizable). Let $\mathscr{C}_{ij}$ denote the pre-specified cost of establishing the transmission link k_{ij}, for any $i, j \in \bar{\nu}$. The implementation cost associated with any set $\mathbf{K}$ is clearly equal to the sum of the costs of its components. Note that in the case when a communication link cannot exist (due to the reasons pointed out earlier such as the shadow phenomenon), its associated cost is infinity. The problems explained below will be addressed in the next section:

- Since any class of structurally constrained controllers is useful only when the system is stabilizable with respect to that class, it is first desired to characterize all classes of LTI structurally constrained controllers with respect to which the system $\mathscr{S}$ has no DOFMs.
- The second objective is to seek the class(es) of LTI structurally constrained controllers (among the ones characterized by addressing the above problem) whose corresponding implementation cost is minimum.

6.4 Main results

The following procedures are essential in developing the main results of this chapter.

Procedure 1 ([7]) *Corresponding to the mode σ_i, $i \in \bar{n} := \{1,2,...,n\}$, construct a bipartite graph $\mathscr{G}_i$ with two sets $\mathscr{V}$ (set 1) and $\bar{\mathscr{V}}$ (set 2). Put ν vertices in each of these sets and label them as $1,2,\ldots,\nu$. For any $\lambda_1,\lambda_2 \in \bar{\nu}$, carry out the steps given below:*

- *Connect the vertex λ_1 of the set $\mathscr{V}$ to the vertex λ_2 of the set $\bar{\mathscr{V}}$ if and only if the following equation holds:*

$$C_{\lambda_1} \times diag\left(\left[\frac{1}{\sigma_1-\sigma_i},\ldots,\frac{1}{\sigma_{i-1}-\sigma_i},0,\right.\right.\\ \left.\left.\frac{1}{\sigma_{i+1}-\sigma_i},\ldots,\frac{1}{\sigma_n-\sigma_i}\right]\right) \times B_{\lambda_2} - D_{\lambda_1\lambda_2} = 0 \tag{6.6}$$

- *Mark the vertex λ_1 of the set $\mathscr{V}$ if the ith column of the matrix C_{λ_1} is a zero vector. Likewise, mark the vertex λ_2 of the set $\bar{\mathscr{V}}$ if the ith row of the matrix B_{λ_2} is a zero vector.*

Procedure 2 *For a given control interaction set $\mathbf{K} = \{k_{p_1q_1}, k_{p_2q_2}, ..., k_{p_\alpha q_\alpha}\}$ and any $i \in \bar{n}$, construct a bipartite graph $\mathscr{G}_i(\mathbf{K})$ with two sets of α vertices. Label the vertices in set 1 with $q_1,q_2,...,q_\alpha$ and the vertices in set 2 with $p_1,p_2,...,p_\alpha$. Mark all vertices in set 1 of $\mathscr{G}_i(\mathbf{K})$ whose corresponding vertices in set 1 of $\mathscr{G}_i$ (the ones with the same labels) are marked. Mark the vertices of set 2 in a similar fashion. Moreover, connect two vertices of set 1 and set 2 of $\mathscr{G}_i(\mathbf{K})$ if the vertices with the same labels in $\mathscr{G}_i$ are connected to each other.*

Since $q_1,q_2,...,q_\alpha$ are not necessarily distinct numbers, the labels in set 1 of the graph $\mathscr{G}_i(\mathbf{K})$ formed in Procedure 2 can be recurrent. The same argument holds true for its second set of vertices. It can be observed that $\mathscr{G}_i$ acts as a look-up table for constructing the graph $\mathscr{G}_i(\mathbf{K})$. As an example to clarify this point, assume the graph $\mathscr{G}_1$ to be the one depicted in Figure 6.1. By considering $\mathbf{K}$ as $\{k_{11},k_{22},k_{13}\}$, the graph $\mathscr{G}_1(\mathbf{K})$ can be obtained easily in terms of $\mathscr{G}_1$ as shown in Figure 6.2.

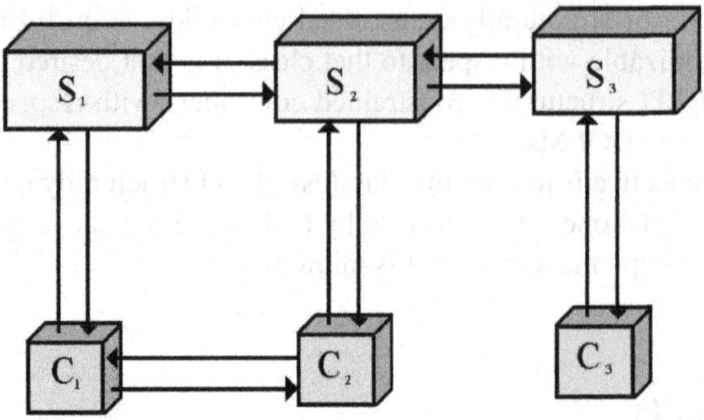

Fig. 6.1 The graph $\mathscr{G}_1$ for a certain system.

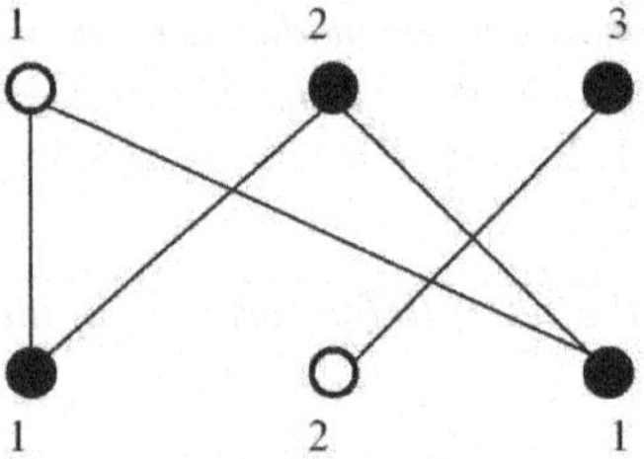

Fig. 6.2 The graph $\mathscr{G}_1(\mathbf{K})$ corresponding to the set $\mathbf{K} = \{k_{11}, k_{22}, k_{13}\}$ and the system with the graph $\mathscr{G}_1$ sketched in Figure 6.1.

Lemma 1 *The mode σ_i, $i \in \bar{n}$, is a DOFM of the system $\mathscr{S}$ with respect to the set $\mathbf{K} = \{k_{p_1q_1}, k_{p_2q_2}, ..., k_{p_\alpha q_\alpha}\}$ if and only if its corresponding graph $\mathscr{G}_i(\mathbf{K})$ satisfies any of the following properties:*

i) All vertices in set 1 of $\mathscr{G}_i(\mathbf{K})$ are marked.
ii) All vertices in set 2 of $\mathscr{G}_i(\mathbf{K})$ are marked.
iii) $\mathscr{G}_i(\mathbf{K})$ includes a complete bipartite subgraph for which both of the conditions given below hold:

- *All its vertices (in both sets) are marked.*
- *For any $j \in \{1,2,...,\alpha\}$, the jth vertex of either set 1 or set 2 of $\mathscr{G}_i(\mathbf{K})$ is included in the subgraph.*

Proof: The proof can be derived straightforwardly from the results of [7] and [16]. The details are omitted here (see Theorem 1 in [18] for a similar result). ∎

Definition 2 *The control interaction set $\mathbf{K} = \{k_{p_1q_1}, k_{p_2q_2}, ..., k_{p_\alpha q_\alpha}\}$ is said to be minimal w.r.t. σ_i, $i \in \bar{n}$, if and only if σ_i is not a DOFM of the system $\mathscr{S}$ w.r.t. $\mathbf{K}$, while it is a DOFM w.r.t. $\mathbf{K} - \{k_{p_jq_j}\}$ for any $j \in \{1,2,...,\alpha\}$.*

The important property of a minimal set is that it has no redundant transmission link, in the sense that all transmission links in the set contribute to the stabilizability of the system. As an example, consider again the system whose corresponding graph $\mathscr{G}_1$ is depicted in Figure 6.1. It is easy to verify that $\{k_{11}, k_{22}, k_{13}\}$ is not a minimal set for this system, while $\{k_{11}, k_{23}\}$ is a minimal one.

Definition 3 *A subgraph of the graph $\mathscr{G}_i$, $i \in \bar{n}$, is said to be maximal if:*

i) It is a complete bipartite graph.
ii) All of its vertices in both sets are marked.
iii) The properties (i) and (ii) given above will not hold if any new vertex is added to the subgraph from the graph $\mathscr{G}_i$.
iv) It includes at least one edge.

Using a proper combinatorial algorithm, these subgraphs of the graph $\mathscr{G}_i$ can be easily identified (analogously to the algorithms developed for finding the complete bipartite subgraphs with maximum number of edges). Denote such subgraphs with $\tilde{\mathscr{G}}^i_1, \tilde{\mathscr{G}}^i_2, ..., \tilde{\mathscr{G}}^i_{w_i}$, for any $i \in \bar{n}$. Moreover, denote set 1 and set 2 of the vertices of the graph $\tilde{\mathscr{G}}^i_j$ with $\tilde{\mathscr{V}}^i_j$ and $\bar{\tilde{\mathscr{V}}}^i_j$, respectively, for any $i \in \bar{n}$ and $j \in \{1, 2, ..., w_i\}$.

To clarify Definition 3, consider $\mathscr{G}_1$ as the graph sketched in Figure 6.3. This graph has two maximal subgraphs with the following sets of vertices:

- $\tilde{\mathscr{V}}^1_1 = \{1\}$ and $\bar{\tilde{\mathscr{V}}}^1_1 = \{2, 3\}$.
- $\tilde{\mathscr{V}}^1_2 = \{1, 2\}$ and $\bar{\tilde{\mathscr{V}}}^1_2 = \{2\}$.

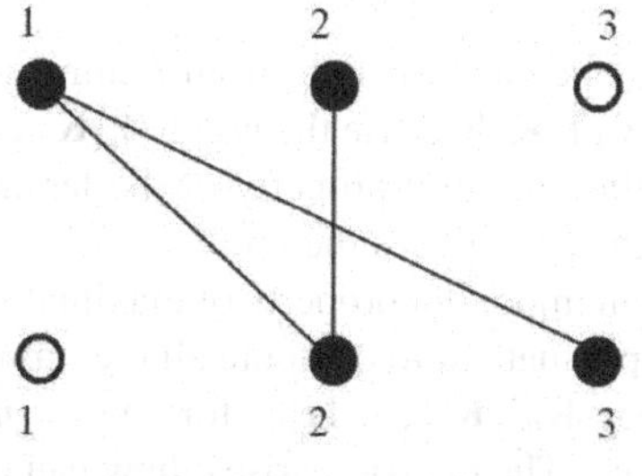

Fig. 6.3 A graph $\mathscr{G}_1$ with two maximal subgraphs.

Theorem 1 *Assume that the control interaction set $\mathbf{K} = \{k_{p_1q_1}, k_{p_2q_2}, ..., k_{p_\alpha q_\alpha}\}$ is minimal w.r.t. σ_i, $i \in \bar{n}$. Then, α is less than or equal to $w_i + 2$.*

Proof: From the definition of a minimal set, the mode σ_i is a DOFM of the system $\mathscr{S}$ w.r.t. $\mathbf{K}^j := \mathbf{K} - \{k_{p_jq_j}\}$ for any $j \in \{1, 2, ..., \alpha\}$. Hence, at least one of the properties (i), (ii) or (iii) mentioned in Lemma 1 is satisfied for the graph $\mathscr{G}_i(\mathbf{K}^j)$ (introduced in Procedure 2) for any $j \in \{1, 2, ..., \alpha\}$. Proof of the theorem will be provided now by contradiction. Therefore, assume that $\alpha > w_i + 2$.

Suppose that there are two distinct numbers $j_1, j_2 \in \{1,2,...,\alpha\}$ so that property (i) of Lemma 1 is met for both of the graphs $\mathscr{G}_i(\mathbf{K}^{j_1})$ and $\mathscr{G}_i(\mathbf{K}^{j_2})$. Hence, all the vertices in set 1 of the graph $\mathscr{G}_i(\mathbf{K})$ are marked. This implies that the system $\mathscr{S}$ has a DOFM w.r.t the set $\mathbf{K}$ (by virtue of Lemma 1). This contradiction means that condition (i) of Lemma 1 can be true for at most one of the graphs $\mathscr{G}_i(\mathbf{K}^1), \mathscr{G}_i(\mathbf{K}^2), ..., \mathscr{G}_i(\mathbf{K}^\alpha)$. The same argument can be made for condition (ii) of Lemma 1. Therefore, condition (iii) of Lemma 1 is true for more than w_i graphs of $\mathscr{G}_i(\mathbf{K}^1), \mathscr{G}_i(\mathbf{K}^2), ..., \mathscr{G}_i(\mathbf{K}^\alpha)$ (note that $\alpha - 2 > w_i$). With no loss of generality, assume that each of the graphs $\mathscr{G}_i(\mathbf{K}^1), \mathscr{G}_i(\mathbf{K}^2), ..., \mathscr{G}_i(\mathbf{K}^\lambda)$ has a complete bipartite subgraph satisfying property (iii) of Lemma 1, where $\lambda = w_i + 1$. Every of these complete bipartite subgraphs is also a subgraph of one of the graphs $\tilde{\mathscr{G}}^i_1, \tilde{\mathscr{G}}^i_2, ..., \tilde{\mathscr{G}}^i_{w_i}$. Since there are λ complete bipartite subgraphs while the number of these graphs is less than λ (note that $w_i < \lambda$), it can be concluded from Dirichlet's Principle that the subgraphs of two of the graphs $\mathscr{G}_i(\mathbf{K}^1), \mathscr{G}_i(\mathbf{K}^2), ..., \mathscr{G}_i(\mathbf{K}^\lambda)$ correspond to the same maximal graph. Without any loss of generality, assume that $\mathscr{G}_i(\mathbf{K}^1)$ and $\mathscr{G}_i(\mathbf{K}^2)$ both have complete bipartite subgraphs with the properties mentioned in condition (iii) of Lemma 1, which are included in $\tilde{\mathscr{G}}^i_1$. Thus:

$$q_\beta \subseteqq \tilde{\mathscr{V}}^i_1 \text{ or } p_\beta \subseteqq \bar{\tilde{\mathscr{V}}}^i_1, \ \forall \beta \in \{2,3,4,...,\alpha\} \tag{6.7}$$

and:

$$q_\beta \subseteqq \tilde{\mathscr{V}}^i_1 \text{ or } p_\beta \subseteqq \bar{\tilde{\mathscr{V}}}^i_1, \ \forall \beta \in \{1,3,4,...,\alpha\} \tag{6.8}$$

It follows from (6.7) and (6.8) that:

$$q_\beta \subseteqq \tilde{\mathscr{V}}^i_1 \text{ or } p_\beta \subseteqq \bar{\tilde{\mathscr{V}}}^i_1, \ \forall \beta \in \{1,2,3,4,...,\alpha\} \tag{6.9}$$

Using the above relation, one can conclude from Lemma 1 that the mode σ_i is a DOFM of the system $\mathscr{S}$ w.r.t. $\mathbf{K}$, because the graph $\mathscr{G}_i(\mathbf{K})$ encompasses a complete bipartite subgraph satisfying the third property of the lemma. This contradicts the initial assumption, and hence completes the proof. ■

Theorem 1 introduces an important property of maximal sets. Moreover, its proof implicitly proposes a simple method to compute all the minimal sets corresponding to the mode σ_i, denoted by $\mathbf{K}_m^{i,1}, \mathbf{K}_m^{i,2}, ..., \mathbf{K}_m^{i,z_i}$, for any $i \in \bar{n}$. Assume for now that all of these sets are attained. The question arises: how can one obtain a set $\mathbf{K}$ such that the system $\mathscr{S}$ has no DOFMs with respect to the control interaction set $\mathbf{K}$? This question will be answered in the next corollary.

Corollary 1 *For any control interaction set* $\mathbf{K}$*, the system* $\mathscr{S}$ *has no DOFMs w.r.t.* $\mathbf{K}$ *if and only if there exist distinct integers* $j_1, j_2, ..., j_n$ *with the following property:*

$$(\mathbf{K}_m^{1,j_1} \cup \mathbf{K}_m^{2,j_2} \cup \cdots \cup \mathbf{K}_m^{n,j_n}) \subseteq \mathbf{K} \tag{6.10}$$

where $1 \le j_i \le z_i$*, for any* $i \in \bar{n}$.

Proof: The proof is straightforward, and is omitted here. ■

Remark 1 *Since the minimal sets* $\mathbf{K}_m^{i,1}, \mathbf{K}_m^{i,2}, ..., \mathbf{K}_m^{i,z_i}$, $i \in \bar{n}$, *have already been identified, Corollary 1 can be utilized to attain all the control structures with respect to which the system has no fixed modes. For any control interaction set* $\mathbf{K}$ *obtained, one can take advantage of the method given in [16] to design a LTI structurally constrained controller complying with* $\mathbf{K}$ *so that all the modes of the system are placed at any desirable locations.*

6.4.1 An efficient algorithm to obtain the optimal control interaction set(s)

It is evident that the implementation cost corresponding to a control interaction set $\mathbf{K} = \{k_{p_1q_1}, k_{p_2q_2}, ..., k_{p_\alpha q_\alpha}\}$ is equal to $\sum_{i=1}^{\alpha} \mathscr{C}_{p_iq_i}$. The objective here is to obtain all the optimal control interaction set(s), i.e. the ones with the properties that not only does the system $\mathscr{S}$ have no DOFMs with respect to them, but their corresponding cost is also minimum.

Consider a control interaction set $\mathbf{K}$. It can be inferred from Corollary 1 that if this set is optimal, then there exist distinct integers $j_1, j_2, ..., j_n$ such that

$$\mathbf{K} = \mathbf{K}_m^{1,j_1} \cup \mathbf{K}_m^{2,j_2} \cup \cdots \cup \mathbf{K}_m^{n,j_n} \tag{6.11}$$

Now, draw a table $\mathscr{T}$ with n columns so that the minimal sets $\mathbf{K}_m^{i,1}, \mathbf{K}_m^{i,2}, ..., \mathbf{K}_m^{i,z_i}$ are placed in the ith column of $\mathscr{T}$ in an arbitrary order, for any $i \in \{1, 2, ..., n\}$. As a result of the equation (6.11), the most straightforward way to obtain the optimal control interaction set(s) is to consider all the possible ways that n sets can be chosen from different columns of $\mathscr{T}$, and for each of the selections, to compute the cost associated with the union of the chosen sets to observe which one has the least cost. This would result in $z_1 \times z_2 \times \cdots \times z_n$ combinatorial operations. A method will be proposed next to significantly reduce this number, in general.

To cast light on the idea here, let start with a particular case. Assume that $z_1 = z_2 = \cdots = z_n = z$, and that $\mathbf{K}_m^{2,1} \subseteq \mathbf{K}_m^{1,1}$. The latter relation means that not only the mode σ_1, but also the mode σ_2 is movable with respect to the interaction set $\mathbf{K}_m^{1,1}$. An indication of the fact is that if the set $\mathbf{K}_m^{1,1}$ is chosen from the first column of the table $\mathscr{T}$, there is no need to opt a set from the second column, as it just incurs cost. Hence, in this case, the number of ways in which $\mathbf{K}_m^{1,1}$ is selected from column 1 is equal to z^{n-2}, rather than z^{n-1}. Consequently, if it is known that the relation $\mathbf{K}_m^{2,1} \subseteq \mathbf{K}_m^{1,1}$ holds, computing $z^{n-2}(z-1)$ combinations is obviated. Typically, there are several such relations (not just one), which makes the method computationally more efficient. Consequently, it is very important to find out the relations such as $\mathbf{K}_m^{2,1} \subseteq \mathbf{K}_m^{1,1}$. This can be done by carrying out $\binom{nz}{2}$ operations, where every operation considers two sets from different columns of the table $\mathscr{T}$ and verifies whether one of them is a subset of the other one. It is worth mentioning that since $\binom{nz}{2}$ is much smaller than $z^{n-2}(z-1)$ in general, this "so-called" pre-processing which aims at finding the desired relations is quite advantageous.

Attribute a set $\mathbf{T}^{i,j}$ to each minimal set $\mathbf{K}_m^{i,j}$, for any $i \in \bar{n}$, $j \in \{1,2,...,z_i\}$, such that the set $\mathbf{T}^{i,j}$ contains a number r if and only if at least one of the sets in the rth column of the table is a subset of $\mathbf{K}_m^{i,j}$. As pointed out earlier in a particular case, the sets $\mathbf{T}^{i,j}$'s can be constructed by carrying out $\binom{z_1+z_2+\cdots+z_n}{2}$ operations. The aforementioned discussion leads to the following corollary.

Corollary 2 *Consider a control interaction set* $\mathbf{K}$. *The system* $\mathscr{S}$ *has no DOFMs w.r.t.* $\mathbf{K}$ *if and only if there exist integers* $i_1, i_2, ..., i_j$ *and* $i'_1, i'_2, ..., i'_j$ $(j \leq n)$ *such that:*

$$\left(\mathbf{K}_m^{i_1,i'_1} \cup \mathbf{K}_m^{i_2,i'_2} \cup \cdots \cup \mathbf{K}_m^{i_j,i'_j}\right) \subseteq \mathbf{K} \tag{6.12}$$

where:

$$\mathbf{T}^{i_1,i'_1} \cup \mathbf{T}^{i_2,i'_2} \cup \cdots \cup \mathbf{T}^{i_j,i'_j} = \{1,2,...,n\} - \{i_1,i_2,...,i_j\} \tag{6.13}$$

Moreover, if the relation (6.12) turns to an equality, i.e.:

$$\mathbf{K}_m^{i_1,i'_1} \cup \mathbf{K}_m^{i_2,i'_2} \cup \cdots \cup \mathbf{K}_m^{i_j,i'_j} = \mathbf{K} \tag{6.14}$$

then $\mathbf{K}$ *is a cost optimal candidate.*

The main advantage of the relation (6.14) over (6.11) to obtain the optimal control interaction set $\mathbf{K}$ is that j can be noticeably less than n. Note that in the case when j is solely one unit less than n, as asserted earlier, there is still a significant saving in the computation time. At this point, a simple algorithm can be devised to obtain all the sets $\mathbf{K}$ representable as (6.14).

Remark 2 *In this chapter, a mode is considered as being either fixed or movable. In other words, the problem is formulated by giving a binary status to each mode, in terms of being or not being a DOFM. However, a mode which is not a DOFM, can be very close to being a DOFM. In this case, the input energy to displace this mode can be undesirably huge [19]. This can cause important practical problems such as input saturation. In order to take this grave issue into consideration, one can define an inherent cost, aside from the implementation cost. This cost should reflect how flexible the modes of the system are. For instance, the notion of approximate decentralized fixed modes (ADFM) introduced in [20] can be used for this purpose. In that case, in the last stage where a control interaction set is to chosen from the possible sets resulted from Corollary 2, both of the implementation cost and the inherent cost should be considered.*

6.5 Numerical example

Consider a system $\mathscr{S}$ consisting of three single-input single-output (SISO) subsystems with the following decoupled state-space matrices:

$$
\begin{aligned}
A &= \begin{bmatrix} 1 & 0 & 0 \\ 0 & 2 & 0 \\ 0 & 0 & 3 \end{bmatrix}, \quad B_1 = \begin{bmatrix} 1 \\ 1 \\ 0 \end{bmatrix}, \quad B_2 = \begin{bmatrix} 0 \\ 3 \\ 5 \end{bmatrix}, \\
B_3 &= \begin{bmatrix} 0 \\ 0 \\ 6 \end{bmatrix}, \quad C_1 = \begin{bmatrix} 0 & 6 & 1 \end{bmatrix} \\
C_2 &= \begin{bmatrix} 0 & 0 & 4 \end{bmatrix}, \quad C_3 = \begin{bmatrix} 10 & 3 & 0 \end{bmatrix} \\
D_{11} &= -1, \quad D_{12} = 23, \quad D_{13} = 3, \\
D_{21} &= -3, \quad D_{22} = 20, \quad D_{23} = 10, \\
D_{31} &= -15, \quad D_{32} = 5, \quad D_{33} = -8,
\end{aligned} \tag{6.15}
$$

It is desired now to characterize all control interaction sets with respect to which the system $\mathscr{S}$ has no DOFMs. The graphs $\mathscr{G}_1$, $\mathscr{G}_2$ and $\mathscr{G}_3$ corresponding to $\sigma_1 = 1$, $\sigma_2 = 2$ and $\sigma_3 = 3$ are depicted in Figures 6.4, 6.5 and 6.6, respectively. The

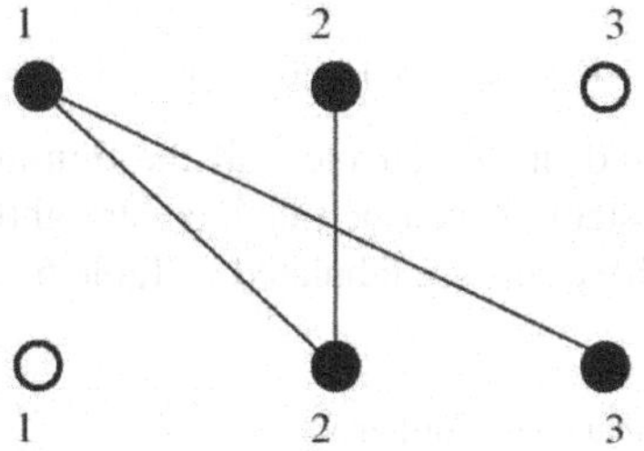

Fig. 6.4 The graph $\mathscr{G}_1$ corresponding to the mode $\sigma_1 = 1$ of the system given in (6.15).

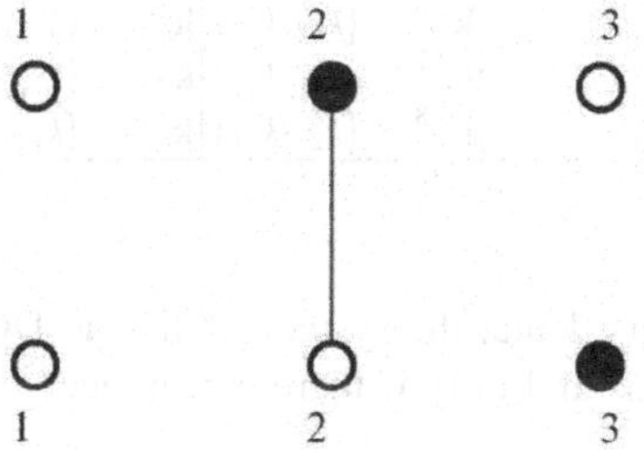

Fig. 6.5 The graph $\mathscr{G}_2$ corresponding to the mode $\sigma_2 = 2$ of the system given in (6.15).

graph $\mathscr{G}_1$ has two maximal subgraphs with the following sets of vertices:

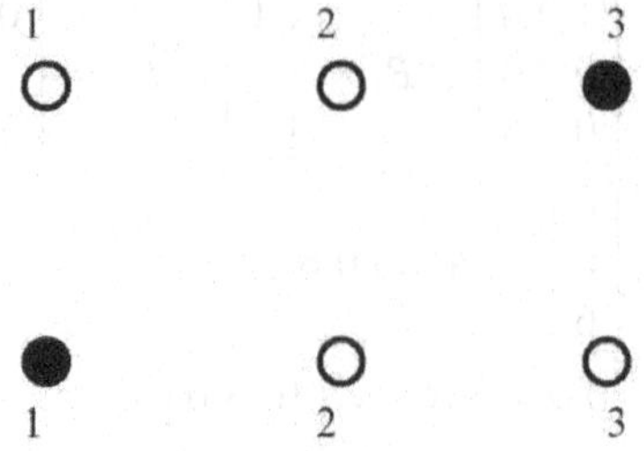

Fig. 6.6 The graph $\mathscr{G}_3$ corresponding to the mode $\sigma_3 = 3$ of the system given in (6.15).

$$\begin{aligned} \tilde{\mathscr{V}}_1^1 &= \{1\}, \quad \bar{\tilde{\mathscr{V}}}_1^1 = \{2,3\} \\ \tilde{\mathscr{V}}_2^1 &= \{1,2\}, \quad \bar{\tilde{\mathscr{V}}}_2^1 = \{2\} \end{aligned} \tag{6.16}$$

Analogously, it can be easily verified that the graphs $\mathscr{G}_2$ and $\mathscr{G}_3$ have no maximal subgraphs. Therefore:

$$w_1 = 2, \quad w_2 = 0, \quad w_3 = 0 \tag{6.17}$$

Using the method proposed in this chapter, all the minimal sets with respect to different modes of the system $\mathscr{S}$ can be obtained straightforwardly based on the maximal graphs found. These sets are tabulated in Table 6.1.

Table 6.1 Minimal sets corresponding to different modes

$\sigma_1 = 1$	$\sigma_2 = 2$	$\sigma_3 = 3$
$\mathbf{K}_m^{1,1} = \{k_{13}\}$	$\mathbf{K}_m^{2,1} = \{k_{11}\}$	$\mathbf{K}_m^{3,1} = \{k_{21}\}$
$\mathbf{K}_m^{1,2} = \{k_{12}, k_{33}\}$	$\mathbf{K}_m^{2,2} = \{k_{21}\}$	$\mathbf{K}_m^{3,2} = \{k_{31}\}$
	$\mathbf{K}_m^{2,3} = \{k_{13}\}$	$\mathbf{K}_m^{3,3} = \{k_{22}\}$
	$\mathbf{K}_m^{2,4} = \{k_{23}\}$	$\mathbf{K}_m^{3,4} = \{k_{32}\}$
	$\mathbf{K}_m^{2,5} = \{k_{22}, k_{33}\}$	$\mathbf{K}_m^{3,5} = \{k_{12}, k_{23}\}$
	$\mathbf{K}_m^{2,6} = \{k_{12}, k_{33}\}$	$\mathbf{K}_m^{3,6} = \{k_{11}, k_{23}\}$
	$\mathbf{K}_m^{2,7} = \{k_{31}, k_{12}\}$	$\mathbf{K}_m^{3,7} = \{k_{11}, k_{33}\}$
	$\mathbf{K}_m^{2,8} = \{k_{22}, k_{31}\}$	$\mathbf{K}_m^{3,8} = \{k_{12}, k_{33}\}$

It results from Corollary 1 that the system $\mathscr{S}$ has no DOFMs with respect to a control interaction set $\mathbf{K}$ if and only if there exit integers $\gamma_1 \in \{1,2\}$ and $\gamma_2, \gamma_3 \in \{1,2,...,8\}$ such that:

$$(\mathbf{K}_m^{1,\gamma_1} \cup \mathbf{K}_m^{2,\gamma_2} \cup \mathbf{K}_m^{3,\gamma_3}) \subseteq \mathbf{K} \tag{6.18}$$

As stated in Subsection A of Section III, in order to characterize the desired sets $\mathbf{K}$, all possible ways of choosing three sets from different columns of Table 6.1 should be considered and for each of them, their corresponding union must be computed. This results in $2 \times 8 \times 8 = 128$ combinations. Nonetheless, Corollary 2 can be exploited to diminish the number of possible combinations. To this end, one may take note of the fact that for any choice of $\mathbf{K}_m^{1,1}$ in column 1 of the table, there is no need to choose a set from column 2 (because of the relation $\mathbf{K}_m^{1,1} = \mathbf{K}_m^{2,3}$). Moreover, it can be deduced from the relation $\mathbf{K}_m^{1,2} = \mathbf{K}_m^{2,6} = \mathbf{K}_m^{3,8}$ that if $\mathbf{K}_m^{1,2}$ is chosen from column 1, no sets are needed to be chosen from the remaining columns. This exposition is indeed the interpretation of Corollary 2. Hence, it suffices to merely consider the union of the first set in column 1 with all sets in column 3 as well as the second set of column 1 itself. In other words, 9 combinations are required to be constructed using Corollary 2, while Corollary 1 requires 128. Obtaining the 9 resultant sets and eliminating the identical ones will lead to the fact that the system has no DOFM with respect to a control interaction set $\mathbf{K}$ if and only if $\mathbf{K}$ includes at least one of the 8 sets given below:

$$\begin{array}{l} \{k_{13},k_{21}\},\ \{k_{13},k_{31}\},\ \{k_{13},k_{22}\},\ \{k_{13},k_{32}\}, \\ \{k_{13},k_{12},k_{23}\},\ \{k_{13},k_{11},k_{23}\},\ \{k_{13},k_{11},k_{33}\}, \\ \{k_{12},k_{33}\} \end{array} \tag{6.19}$$

It is worth mentioning that there exist $2^{(3^2)} - 1 = 511$ classes of structurally constrained controllers for this example. However, the desired ones are characterized here as only 9 minimal sets.

Now, it is desired to obtain the *cost optimal* control interaction set(s). Assume that $\mathscr{C}_{ij} = 1,\ i,j \in \{1,2,3\}$, which implies that establishing any of the transmission links would incur the same cost. As discussed earlier, among the sets given in (6.19), the ones whose corresponding costs are minimum should be identified. In this case, the five sets which have only 2 elements are the cost optimal ones with the minimum cost 2.

6.6 Summary

This chapter aims to obtain all classes of LTI structurally constrained controllers with respect to which a given interconnected system has no fixed modes. To this end, the notions of maximal graph and minimal set are exploited to formulate the problem in the graph theory framework. A cost is then allocated for establishing a communication link between any pair of controllers. An algorithm is subsequently developed to identify the stabilizing control structures with the minimum implementation cost. The significance of this contribution is illustrated in a numerical example. This work takes advantage of the recent results presented in the literature to handle similar problems.

References

1. J. Lavaei and A. G. Aghdam, "Decentralized control design for interconnected systems based on a centralized reference controller," in *Proceedings of 45th IEEE Conference on Decision and Control*, San Diego, CA, 2006.
2. J. Lavaei, A. Momeni and A. G. Aghdam, "A model predictive decentralized control scheme with reduced communication requirement for spacecraft formation", *IEEE Transactions on Control Systems Technology*, vol. 16, no. 2, pp. 268-278, 2008.
3. R. S. Smith and F. Y. Hadaegh, "Control topologies for deep space formation flying spacecraft," in *Proceedings of 2002 American Control. Conference*, Anchorage, AK, 2002.
4. J. Lavaei and A. G. Aghdam, "Optimal periodic feedback design for continuous-time LTI systems with constrained control structure," *International Journal of Control*, vol. 80, no. 2, pp. 220-230, 2007.
5. E. J. Davison and T. N. Chang, "Decentralized stabilization and pole assignment for general proper systems," *IEEE Transactions on Automatic Control*, vol. 35, no. 6, pp. 652-664, 1990.
6. D. D. Šiljak, Decentralized control of complex systems, Cambridge: Academic Press, 1991.
7. J. Lavaei and A. G. Aghdam, "Characterization of decentralized and quotient fixed modes via graph theory," in *Proceedings of 2007 American Control Conference*, New York, NY, 2007.
8. M. S. Ravi, J. Rosenthal and X. A. Wang, "On decentralized dynamic pole placement and feedback stabilization," *IEEE IEEE Transactions on Automatic Control*, vol. 40, no. 9, pp. 1603-1614, 1995.
9. J. Lavaei and A. G. Aghdam, "Elimination of fixed modes by means of high-performance constrained periodic control," in *Proceedings of 45th IEEE Conference on Decision and Control*, San Diego, CA, 2006.
10. B. A. Movsichoff, C. M. Lagoa and H. Che, "Decentralized optimal traffic engineering in connectionless networks," *IEEE Journal on Selected Areas in Communications*, vol. 23, no. 2, pp. 293-303, 2005.
11. D. M. Stipanovi, G. Inalhan, R. Teo and C. J. Tomlin, "Decentralized overlapping control of a formation of unmanned aerial vehicles," *Automatica*, vol. 40 , no. 8, pp. 1285-1296, 2004.
12. A. I. Zecevic and D. D. Šiljak, "A new approach to control design with overlapping information structure constraints," *Automatica*, vol. 41, no. 2, pp. 265-272, 2005.
13. A. Iftar, "Overlapping decentralized dynamic optimal control," *International Journal of Control*, vol. 58, no. 1, pp. 187-209, 1993.
14. S. H. Wang and E. J. Davison, "On the stabilization of decentralized control systems," *IEEE Transactions on Automatic Control*, vol. 18, no. 5, pp. 473-478, 1973.
15. Z. Gong and M. Aldeen, "Stabilization of decentralized control systems," *Journal of Mathematical Systems, Estimation, and Control*, vol. 7, no. 1, pp. 1-16, 1997.
16. J. Lavaei and A. G. Aghdam, "A necessary and sufficient condition for the existence of a LTI stabilizing decentralized overlapping controller," in *Proceedings of 45th IEEE Conference on Decision and Control*, San Diego, CA, 2006.
17. J. Lavaei and A. G. Aghdam, "On structurally constrained control design with a prespecified form," in *Proceedings of 2007 American Control Conference*, New York, NY, 2007.
18. S. Sojoudi, J. Lavaei and A. G. Aghdam, "Optimal information flow structure for control of interconnected systems," in *Proceedings of 2007 American Control Conference*, New York, NY, 2007.
19. A. G. Aghdam and E. J. Davison, "Digital control design for systems with approximate decentralized fixed modes," in *Proceedings of 45th IEEE Conference on Decision and Control*, San Diego, CA, 2006.
20. A. F. Vaz and E. J. Davison, "On the quantitative characterization of approximate decentralized fixed modes using transmission zeros," *Mathematics of Control Signal, and Systems*, vol. 2, pp. 287-302, 1989.

Chapter 7
LQ Decentralized Controllers with Disturbance Rejection Property for Hierarchical Systems

7.1 Introduction

In the control literature, an interconnected system is often referred to a system with a collection of interacting subsystems [1]. In terms of the interaction topology between the subsystems, the class of *hierarchical* interconnected systems has drawn special attention in recent publications due to its broad applications such as formation flying, underwater vehicles, automated highway, robotics, satellite constellation, etc., which have leader-follower structures or structures with virtual leaders [2, 3, 4, 5, 6]. It is shown in [2] that even if a continuous-time interconnected system does not have a hierarchical structure, under certain conditions its discrete-time equivalent model can be transformed to a hierarchical form. For such a system, it is normally desired to design a set of local controllers corresponding to the individual subsystems, which partially exchange their information [4, 7]. This demand is originated from some practical limitations concerning, for instance, the geographical distribution of the subsystems or the computational complexity associated with a centralized controller [8]. The case when these local controllers operate independently (i.e., they do not interact with each other), is referred to as decentralized feedback control [9, 10, 11].

Various aspects of the decentralized control theory have been extensively investigated in the past few decades. The papers [10, 12, 13, 14] study the decentralized stabilizability of a system by using the notions of decentralized fixed modes and quotient fixed modes. Several approaches are proposed in the literature to solve the pole-placement problem by means of decentralized controllers [15, 16]. Furthermore, high-performance decentralized control design techniques have been investigated in [18, 19].

Since the real-world systems are usually vulnerable to external disturbances, the controller being designed for a hierarchical interconnected system is desired to satisfy the following properties:

i) The disturbances must be rejected in the steady state.

ii) A predefined H_2 performance index should be minimized to achieve a fast transient response with an acceptable control energy.
iii)The controller to be designed should be decentralized.

There exist a number of works which have addressed the problem of designing a controller satisfying the properties (i) and (iii) given above, and the controller obtained is regarded as [20, 21, 22]. The paper [20] parameterizes all the decentralized controllers which reject the unmeasurable disturbances with known dynamics.

Moreover, the design of a controller which meets the criteria (ii) and (iii) has been studied intensively in several papers. In contrast to the H_2 optimal centralized controller which can be simply obtained from the Riccati equation, the H_2 optimal decentralized control problem involves sophisticated differential/nonconvex matrix equations [23, 24]. As a result, the available techniques often seek a near-optimal solution, rather than a globally optimal one. For instance, a method is proposed in [25], which cuts off all the interconnections between the subsystems and designs local optimal controllers for the isolated subsystems accordingly. The main shortcoming of this approach is that the controller obtained may destabilize the system, in light of neglecting the system's interconnection parameters in the controller design. Another technique for handling the underlying problem is to impose the stringent constraint of staticness on the decentralized controller to be designed [26, 27]. More recently, a method is provided in [8] which decentralizes any given centralized controller of desired performance. As a by-product of the results in [8], it is shown that the decentralized version of the H_2 optimal centralized controller is a H_2 near-optimal decentralized controller. The only requirement of this approach is that the nominal model of the system is known by all control agents; i.e., every local controller must have a belief about the model of the entire system. This idea is further developed in [11] for the flight formation problem in the model predictive control framework. The paper [28], on the other hand, aims to design a controller for which all the aforementioned criteria (i), (ii) and (iii) hold. Since a set of nonlinear equations are derived in [28] for control design, this work cannot tackle the problem in question efficiently.

This chapter presents a novel design strategy to obtain a high-performance decentralized control law for hierarchical interconnected systems, which is able to attenuate the effects of unmeasurable external disturbances with known dynamics. It is assumed that the state of each subsystem is available in its corresponding local output (this is not an unrealistic assumption in many applications such as vehicle formation problems [29]), and that the modeling parameters of the whole system are available (with some error) in any local station. It is to be noted that once a centralized controller is designed to achieve the properties (i) and (ii), its decentralized version (as can be obtained from [8]) does not necessarily maintain the same properties. To bypass this hurdle, a centralized controller is first designed to satisfy some artificial constraints (inspired by the conditions given in [11]). This controller is formulated in terms of LMI and can be obtained straightforwardly. The centralized controller elicited from the LMI problem is subsequently decentralized via the approach presented in [8]. Two issues concerning the designed controller are investigated. First, since the knowledge of each local controller about the whole system is inexact in practice, a procedure is proposed to measure the closeness of the designed decentralized controller to the optimal one in terms of the statistical information on the parameter deviation. This enables the designer to assess the performance of the proposed controller in the real-world environment. This will be demonstrated later in a formation flying example, which shows that the proposed control law outperforms the methods surveyed earlier. Furthermore, the robustness analysis results are presented for the case when the system is subject to perturbation and input delay.

7.2 Problem formulation

Consider a hierarchical interconnected system $\mathscr{S}$, whose ith subsystem $\mathscr{S}_i$, $i \in \bar{\nu} := \{1,2,...,\nu\}$, is represented by:

$$\begin{aligned} \dot{x}_i(t) &= \sum_{j=1}^{i} A_{ij}x_j(t) + B_i u_i(t) + E_i\omega(t) \\ y_i(t) &= C_i x_i(t) \end{aligned} \tag{7.1}$$

where $x_i \in \Re^{n_i}$ and $u_i \in \Re^{m_i}$ are the state and the input of the subsystem $\mathscr{S}_i$, respectively. Furthermore, $y_i \in \Re^{r_i}$ is the output of $\mathscr{S}_i$ to be regulated, and $\omega(t) \in \Re^q$ is the disturbance vector. Assume that the state $x_i(t)$ of the subsystem $\mathscr{S}_i$ is locally available, and that there is no measurement noise, i.e., the measured output in the ith subsystem is equal to $x_i(t)$. Suppose that the disturbance $\omega(t)$ can be expressed as:

$$\begin{aligned} \dot{z}(t) &= \Lambda z(t) \\ \omega(t) &= \mathscr{C} z(t) \end{aligned} \tag{7.2}$$

where the pair $(\mathscr{C}, \Lambda)$ is observable, and $z(0)$ is arbitrary and unknown.

The system $\mathscr{S}$ can be represented as follows:

$$\begin{aligned} \dot{x}(t) &= Ax(t) + Bu(t) + E\omega(t) \\ y(t) &= Cx(t) \end{aligned} \tag{7.3}$$

where:

$$\begin{aligned} &x(t) = \left[x_1(t)^T \; x_2(t)^T \; \cdots \; x_\nu(t)^T\right]^T, \quad u(t) = \left[u_1(t)^T \; u_2(t)^T \; \cdots \; x_\nu(t)^T\right]^T \\ &y(t) = \left[y_1(t)^T \; y_2(t)^T \; \cdots \; y_\nu(t)^T\right]^T, \quad E = \left[E_1^T \; E_2^T \; \cdots \; E_\nu^T\right]^T \\ &B = \mathrm{diag}\left(\left[B_1 \; B_2 \; \cdots \; B_\nu\right]\right), \quad C = \mathrm{diag}\left(\left[C_1 \; C_2 \; \cdots \; C_\nu\right]\right) \end{aligned} \tag{7.4}$$

and A is a $\nu \times \nu$ lower block triangular matrix whose (i, j) block entry is equal to A_{ij}, for any $i, j \in \bar{\nu}$, $j \leq i$. Define now:

$$n := \sum_{i=1}^{\nu} n_i, \quad m := \sum_{i=1}^{\nu} m_i, \quad r := \sum_{i=1}^{\nu} r_i \tag{7.5}$$

Suppose that the initial state $x(0)$ is a random variable with a given mean X_μ and variance X_σ. Define X_0 as:

$$X_0 := \mathscr{E}\{x(0)x(0)^T\} = X_\sigma + X_\mu X_\mu^T \tag{7.6}$$

where $\mathscr{E}\{\cdot\}$ represents the expectation operator. Furthermore, assume that the elements of the matrix E given in (7.3) are arbitrary and unknown. The objective of this chapter is introduced in Problem 1 given below.

Problem 1: Design a decentralized LTI controller K_d (with block diagonal information flow structure [10]), such that the following conditions hold:

i) The state $x(t)$ goes to zero as $t \to \infty$, provided $z(0) = 0$.
ii) The output $y(t)$ approaches zero as $t \to \infty$, regardless of the initial state $z(0)$.
iii) When $z(0)$ is a zero vector, the performance index J corresponding to the closed-loop system is satisfactorily small, where:

$$J := \mathscr{E}\left\{\int_0^\infty \left(x(t)^T Q x(t) + u(t)^T R u(t)\right) dt\right\} \tag{7.7}$$

and where $R \in \Re^{m\times m}$ and $Q \in \Re^{n\times n}$ are positive definite and positive semi-definite matrices, respectively.

It is to be noted that since E is an unknown matrix and can take any arbitrary value, it is desired that the controller K_d be independent of E. The results obtained can be easily extended to the tracking problem in the presence of nonzero reference input, provided it can be expressed similarly to (7.2), with all of its modes located in the closed left half plane. This inclusion can be carried out by defining an augmented system and converting the tracking problem for the original system to a regulation one for the augmented system [30]. The following assumption is made without loss of generality.

Assumption 1: The matrices B_i, C_i and $\mathscr{C}$ satisfy the following relations:

$$\operatorname{rank}(\mathscr{C}) = q, \quad \operatorname{rank}(B_i) = m_i, \quad \operatorname{rank}(C_i) = r_i, \quad i \in \bar{\nu} \tag{7.8}$$

One can easily deduce from the results of [20] that the following two assumptions are required for the existence of the desired controller $\mathscr{K}_d$ under Assumption 1.

Assumption 2: The matrices given below are all full-rank:

$$\begin{bmatrix} A_{ii} - \lambda_j I & B_i \\ C_i & 0 \end{bmatrix}, \quad \forall\, i \in \bar{\nu} \quad \text{and} \quad \forall\, j \in \{1, 2, \ldots, p\} \tag{7.9}$$

where $\lambda_1, \lambda_2, \ldots, \lambda_p$ denote the eigenvalues of Λ. Furthermore, the inequality $m_i \geq r_i$ holds for all $i \in \bar{\nu}$.

Assumption 3: The pair (A_{ii}, B_i) is stabilizable for all $i \in \bar{\nu}$.

7.3 Preliminaries

In this section, it is desired to present the gist of the decentralization procedure given in [8]. Assume for now that $z(0) = 0$, i.e., the system $\mathscr{S}$ is disturbance free. Define the following vectors:

$$\begin{aligned} x^i(t) &= \left[x_1(t)^T \dots x_{i-1}(t)^T \; x_{i+1}(t)^T \dots x_\nu(t)^T \right]^T, \\ u^i(t) &= \left[u_1(t)^T \dots u_{i-1}(t)^T \; u_{i+1}(t)^T \dots u_\nu(t)^T \right]^T, \end{aligned} \tag{7.10}$$

for any $i \in \bar{\nu}$. Consider an arbitrary centralized LTI controller K_c with the following state-space representation:

$$\begin{aligned} \dot{\eta}_c(t) &= \Gamma \eta_c(t) + \Omega x(t) \\ u(t) &= M\eta_c(t) + Nx(t) \end{aligned} \tag{7.11}$$

where $\eta_c \in \Re^\mu$. There exist constant matrices $\Omega^i, \Omega_i, \mathbf{M}^i, \mathbf{M}_i, \mathbf{N}^i, \mathbf{N}_i, \mathbf{N}^{\bar{i}}$ and $\mathbf{N}_{\bar{i}}$, such that the above controller can be expressed in a decomposed representation as follows:

$$\begin{aligned} \dot{\eta}_c(t) &= \Gamma \eta_c(t) + \Omega^i x^i(t) + \Omega_i x_i(t) \\ u^i(t) &= \mathbf{M}^i \eta_c(t) + \mathbf{N}^i x^i(t) + \mathbf{N}_i x_i(t) \\ u_i(t) &= \mathbf{M}_i \eta_c(t) + \mathbf{N}^{\bar{i}} x^i(t) + \mathbf{N}_{\bar{i}} x_i(t) \end{aligned} \tag{7.12}$$

for any $i \in \bar{\nu}$. Similarly, there exist matrices $\mathbf{A}^i, \mathbf{A}^{\bar{i}}, \mathbf{A}_i$ and $\mathbf{B}^i$ (derived from A and B) such that the system $\mathscr{S}$ given in (7.1) can be decomposed as follows:

$$\begin{aligned} \dot{x}^i(t) &= \mathbf{A}^i x^i(t) + \mathbf{A}_i x_i(t) + \mathbf{B}^i u^i(t) \\ \dot{x}_i(t) &= \mathbf{A}^{\bar{i}} x^i(t) + A_{ii} x_i(t) + B_i u_i(t) \end{aligned} \tag{7.13}$$

for any $i \in \bar{\nu}$. Define K_{d_i} as a local controller for the subsystem $\mathscr{S}_i$, $i \in \bar{\nu}$, with the following state-space representation:

$$\begin{aligned} \dot{\eta}_{d_i}(t) &= \begin{bmatrix} \mathbf{A}^i + \mathbf{B}^i \mathbf{N}^i & \mathbf{B}^i \mathbf{M}^i \\ \Omega^i & \Gamma \end{bmatrix} \eta_{d_i}(t) + \begin{bmatrix} \mathbf{A}_i + \mathbf{B}^i \mathbf{N}_i \\ \Omega_i \end{bmatrix} x_i(t) \\ u_i(t) &= \begin{bmatrix} \mathbf{N}^{\bar{i}} & \mathbf{M}_i \end{bmatrix} \eta_{d_i}(t) + \mathbf{N}_{\bar{i}} x_i(t) \end{aligned} \tag{7.14}$$

Define also K_d as a decentralized controller consisting of the local controllers $K_{d_1}, K_{d_2}, \dots, K_{d_\nu}$.

Theorem 1 *[8] Assume that $x(0)$ is a known vector (as opposed to a random variable). The state and the input of the system $\mathscr{S}$ under the centralized controller K_c are the same as those of the system $\mathscr{S}$ under the decentralized controller K_d, if the initial state of the local controller K_{d_i} is chosen as:*

$$\eta_{d_i}(0) = \begin{bmatrix} x^i(0) \\ 0 \end{bmatrix}, \quad i \in \bar{\nu} \tag{7.15}$$

Theorem 1 states that the centralized controller K_c can be transformed to an equivalent decentralized controller K_d, if the initial state $x(0)$ is a known vector and any local controller K_{d_i}, $i \in \bar{\nu}$, exactly knows the initial states of the other subsystems. It is to be noted that these are not realistic assumptions in practice, and thus the result of Theorem 1 cannot be applied to the real-world problems. However, this result will be used later for the development of the main results of this chapter, where the practical limitations are taken into account. As the first step, assume that $x(0)$ is only statistically known, and hence let the following initial state be deployed:

$$\eta_{d_i}(0) = \begin{bmatrix} X_\mu^i \\ 0 \end{bmatrix}, \quad i \in \bar{\nu} \tag{7.16}$$

instead of the one in (7.15). In the sequel, the internal stability of the system $\mathscr{S}$ under the decentralized controller K_d will be investigated.

Definition 1 *Consider the system $\mathscr{S}$ given by (7.1). The modified system $\mathbf{S}^i$, $i \in \bar{\nu}$, is defined to be a system obtained by removing all interconnections going to the i'th subsystem in $\mathscr{S}$. The state-space representation of the modified system $\mathbf{S}^i$ is as follows:*

$$\begin{aligned} \dot{x}(t) &= \tilde{A}^i x(t) + Bu(t) \\ y(t) &= Cx(t) \end{aligned} \tag{7.17}$$

where $\tilde{A}^i$ is derived from A by replacing the first $i-1$ block entries of its i'th block row with zeros. It is to be noted that $\mathbf{S}^1 = \mathscr{S}$.

Definition 2 *Define the isolated subsystem $\mathscr{S}_i$, $i \in \bar{\nu}$, as a system obtained from the subsystem $\mathscr{S}_i$ by eliminating all of its incoming interconnections.*

Theorem 2 *[11] The system $\mathscr{S}$ is internally stable under the controller K_d if and only if the system $\mathbf{S}^i$ is stable under the controller K_c, for all $i \in \bar{\nu}$.*

A centralized servomechanism controller will be given in the next section, which will be used later as a reference to obtain the desired decentralized controller.

7.4 A reference centralized servomechanism controller

To avoid trivial cases, assume with no loss of generality that all of the eigenvalues of Λ lie in the closed right-half plane. It can be concluded from Assumptions 1, 2 and 3, and the results of [31], that there exist three nonunique matrices $\mathscr{B}, \mathscr{M}$ and $\mathscr{N}$, such that any minimum order *centralized* controller satisfying the requirements (i) and (ii) of Problem 1 can be represented by:

$$\dot{\eta}_c(t) = \mathscr{A}\eta_c(t) + \mathscr{B}y(t) \tag{7.18a}$$

$$u(t) = \mathscr{M}\eta_c(t) + \mathscr{N}x(t) \tag{7.18b}$$

where:

$$\mathscr{A} := \mathrm{diag}([\underbrace{\Lambda \ \Lambda \ \cdots \ \Lambda}_{r \text{ times}}]) \tag{7.19}$$

and where $(\mathscr{A}, \mathscr{B})$ is controllable. The objective of this section is to solve the problem introduced below.

Problem 2: Find the matrices $\mathscr{B}, \mathscr{M}$, and $\mathscr{N}$, so that the centralized controller given by (7.18) has the following properties:

i) It satisfies the criteria (i) and (ii) of Problem 1.
ii) It stabilizes all of the systems $\mathbf{S}^2, ..., \mathbf{S}^\nu$.
iii) $\mathscr{B}$ is a block diagonal matrix, and the dimension of its ith block entry is $r_i p \times r_i$, for any $i \in \bar{\nu}$.

The centralized controller satisfying the conditions of Problem 2 will be transformed to a decentralized controller in the next section. It is to be noted that the conditions (ii) and (iii) given above are required in the decentralization procedure, as will be shown subsequently.

Lemma 1 *Problem 2 has a solution, if and only if there exist a block diagonal matrix $\mathscr{B}$, matrices $\mathscr{M}$ and $\mathscr{N}$, and positive definite matrices $P_1, P_2, ..., P_\nu$ with the following properties:*

$$
\begin{bmatrix} \tilde{A}^i & 0 \\ \mathscr{B}C & \mathscr{A} \end{bmatrix}^T P_i + P_i \begin{bmatrix} \tilde{A}^i & 0 \\ \mathscr{B}C & \mathscr{A} \end{bmatrix} - P_i \begin{bmatrix} B \\ 0 \end{bmatrix} R^{-1} \begin{bmatrix} B \\ 0 \end{bmatrix}^T P_i
$$
$$
+ \left(R^{-\frac{1}{2}} \begin{bmatrix} B \\ 0 \end{bmatrix}^T P_i + R^{\frac{1}{2}} [\mathscr{N} \ \mathscr{M}] \right)^T \tag{7.20}
$$
$$
\times \left(R^{-\frac{1}{2}} \begin{bmatrix} B \\ 0 \end{bmatrix}^T P_i + R^{\frac{1}{2}} [\mathscr{N} \ \mathscr{M}] \right) + \begin{bmatrix} \mathscr{Q} & 0 \\ 0 & 0 \end{bmatrix} < 0, \qquad i \in \bar{\nu}
$$

Proof: Substituting (7.18a) into (7.17) results in the augmented system given below:

$$
\begin{bmatrix} \dot{x}(t) \\ \dot{\eta}_c(t) \end{bmatrix} = \begin{bmatrix} \tilde{A}^i & 0 \\ \mathscr{B}C & \mathscr{A} \end{bmatrix} \begin{bmatrix} x(t) \\ \eta_c(t) \end{bmatrix} + \begin{bmatrix} B \\ 0 \end{bmatrix} u(t), \quad i \in \bar{\nu} \tag{7.21}
$$

It is inferred from [31] that the desired controller exists, if and only if there exist a block diagonal matrix $\mathscr{B}$, and matrices $\mathscr{M}$ and $\mathscr{N}$ such that the static controller $u(t) = [\mathscr{N} \ \mathscr{M}] \begin{bmatrix} x(t) \\ \eta_c(t) \end{bmatrix}$ stabilizes all of the augmented systems given by (7.21). Moreover, it follows from [32] that this stabilizability problem is equivalent to the solvability of the matrix inequality problem given in (7.20). ∎

Theorem 3 *Problem 2 has a solution, if and only if there exist block diagonal matrices $\mathscr{B}$ and W, matrices $\mathscr{M}$ and $\mathscr{N}$, and positive definite matrices $P_1, P_2, ..., P_\nu, V_1, V_2, ..., V_\nu$, such that the following matrix inequality problem:*

$$
\begin{bmatrix} \Phi_i & \bar{\Phi}_i \\ \bar{\Phi}_i^T & -I \end{bmatrix} < 0, \quad i \in \bar{\nu} \tag{7.22}
$$

is feasible, where:

$$\begin{aligned}\Phi_i &= \begin{bmatrix} \tilde{A}^i & 0 \\ 0 & \mathscr{A} \end{bmatrix}^T P_i + P_i \begin{bmatrix} \tilde{A}^i & 0 \\ 0 & \mathscr{A} \end{bmatrix} + V_i \left(\begin{bmatrix} B \\ 0 \end{bmatrix} R^{-1} \begin{bmatrix} B \\ 0 \end{bmatrix}^T + I \right) V_i \\ &\quad - P_i \left(\begin{bmatrix} B \\ 0 \end{bmatrix} R^{-1} \begin{bmatrix} B \\ 0 \end{bmatrix}^T + I \right) V_i - V_i \left(\begin{bmatrix} B \\ 0 \end{bmatrix} R^{-1} \begin{bmatrix} B \\ 0 \end{bmatrix}^T + I \right) P_i \\ &\quad + \begin{bmatrix} I \\ 0 \end{bmatrix} C^T \left(W^T W - \mathscr{B}^T W - W^T \mathscr{B} \right) C \begin{bmatrix} I & 0 \end{bmatrix} + \begin{bmatrix} Q & 0 \\ 0 & 0 \end{bmatrix} \\ \bar{\Phi}_i &= \left[\left(R^{-\frac{1}{2}} \begin{bmatrix} B \\ 0 \end{bmatrix}^T P_i + R^{\frac{1}{2}} \begin{bmatrix} \mathscr{N} & \mathscr{M} \end{bmatrix} \right)^T \begin{bmatrix} 0 & 0 \\ \mathscr{B}C & 0 \end{bmatrix}^T + P_i \right], \quad i \in \bar{\nu} \end{aligned} \tag{7.23}$$

Proof of necessity: Assume that Problem 2 has a solution. It can be concluded from Lemma 1 that there exist a block diagonal matrix $\mathscr{B}$, matrices $\mathscr{M}$ and $\mathscr{N}$, and positive definite matrices $P_1, P_2, \ldots, P_\nu$, such that the matrix inequality problem given in (7.20) is feasible. One can easily verify that the matrix inequality problem (7.22) for the matrix variables $\mathscr{B}, \mathscr{M}, \mathscr{N}, W, P_1, \ldots, P_\nu, V_1, \ldots, V_\nu$, where $W = \mathscr{B}$ and $V_i = P_i$, $\forall i \in \bar{\nu}$, is the same as the one expressed by (7.20).

Proof of sufficiency: Suppose that there exist block diagonal matrices $\mathscr{B}$ and W, matrices $\mathscr{M}$ and $\mathscr{N}$, and positive definite matrices $P_1, \ldots, P_\nu, V_1, \ldots, V_\nu$, such that the matrix inequality problem (7.22) is feasible. Applying the Schur complement's formula to (7.22), one can conclude that:

$$\bar{\Phi}_i \bar{\Phi}_i^T + \Phi_i < 0, \quad i \in \bar{\nu} \tag{7.24}$$

On the other hand, it is known that:

$$(P_i - V_i) \left(\begin{bmatrix} B \\ 0 \end{bmatrix} R^{-1} \begin{bmatrix} B \\ 0 \end{bmatrix}^T + I \right) (P_i - V_i) \geq 0, \quad C^T (B - W)^T (B - W) C \geq 0 \tag{7.25}$$

The above inequalities are equivalent to the following ones:

$$\begin{aligned} & V_i \left(\begin{bmatrix} B \\ 0 \end{bmatrix} R^{-1} \begin{bmatrix} B \\ 0 \end{bmatrix}^T + I \right) V_i - V_i \left(\begin{bmatrix} B \\ 0 \end{bmatrix} R^{-1} \begin{bmatrix} B \\ 0 \end{bmatrix}^T + I \right) P_i \\ & - P_i \left(\begin{bmatrix} B \\ 0 \end{bmatrix} R^{-1} \begin{bmatrix} B \\ 0 \end{bmatrix}^T + I \right) V_i \geq - P_i \left(\begin{bmatrix} B \\ 0 \end{bmatrix} R^{-1} \begin{bmatrix} B \\ 0 \end{bmatrix}^T + I \right) P_i \end{aligned} \tag{7.26a}$$

$$C^T W^T W C - C^T \mathscr{B}^T W C - C^T W^T \mathscr{B} C \geq - C^T \mathscr{B}^T \mathscr{B} C \tag{7.26b}$$

The inequalities (7.24), (7.26a) and (7.26b) lead to the following:

$$\begin{aligned}&\bar{\Phi}_i\bar{\Phi}_i^T+\begin{bmatrix}\tilde{A}^i & 0\\ 0 & \mathscr{A}\end{bmatrix}^T P_i+P_i\begin{bmatrix}\tilde{A}^i & 0\\ 0 & \mathscr{A}\end{bmatrix}-P_i\left(\begin{bmatrix}B\\0\end{bmatrix}R^{-1}\begin{bmatrix}B\\0\end{bmatrix}^T+I\right)P_i\\&-\begin{bmatrix}I\\0\end{bmatrix}C^T\mathscr{B}^T\mathscr{B}C\begin{bmatrix}I & 0\end{bmatrix}+\begin{bmatrix}Q & 0\\ 0 & 0\end{bmatrix}<0,\quad i\in\bar{\mathrm{v}}\end{aligned} \tag{7.27}$$

The proof follows from Lemma 1 and from the fact that the expressions in the left sides of the inequalities (7.20) and (7.27) are identical. ■

Remark 1 *It can be easily verified that the matrix inequalities (7.22) turn to be LMIs when V and W are set to be constants.*

Consider now the ith isolated subsystem $\mathscr{S}_i$:

$$\begin{aligned}\dot{x}_i(t)&=A_{ii}x_i(t)+B_iu_i(t)+E_i\omega(t)\\ y_i(t)&=C_ix_i(t)\end{aligned} \tag{7.28}$$

Pursuing the method proposed in [31] and using Assumptions 1, 2 and 3, one can obtain the matrices $\mathscr{B}_i,\mathscr{M}_i$ and $\mathscr{N}_i$, for any $i\in\bar{\mathrm{v}}$, such that the controller:

$$\begin{aligned}\dot{\eta}_{c_i}(t)&=\mathscr{A}_i\eta_{c_i}(t)+\mathscr{B}_iy_i(t)\\ u_i(t)&=\mathscr{M}_i\eta_{c_i}(t)+\mathscr{N}_ix_i(t)\end{aligned} \tag{7.29}$$

attenuates the state $x_i(0)$ of the system given in (7.28) to zero provided $z(0)=0$, and regulates $y_i(t)$ to zero for any arbitrary $z(0)$, where:

$$\mathscr{A}_i:=\operatorname{diag}([\underbrace{\Lambda\ \Lambda\ \cdots\ \Lambda}_{r_i\text{ times}}]) \tag{7.30}$$

Define now the following matrices:

$$\begin{aligned}\mathscr{B}_o&=\operatorname{diag}\left(\left[\mathscr{B}_1\ \mathscr{B}_2\ \cdots\ \mathscr{B}_{\mathrm{v}}\right]\right),\\ \mathscr{M}_o&=\operatorname{diag}\left(\left[\mathscr{M}_1\ \mathscr{M}_2\ \cdots\ \mathscr{M}_{\mathrm{v}}\right]\right),\\ \mathscr{N}_o&=\operatorname{diag}\left(\left[\mathscr{N}_1\ \mathscr{N}_2\ \cdots\ \mathscr{N}_{\mathrm{v}}\right]\right)\end{aligned} \tag{7.31}$$

By considering $\mathscr{B}=\mathscr{B}_0$, $\mathscr{M}=\mathscr{M}_0$, and $\mathscr{N}=\mathscr{N}_0$, it can be easily concluded that the controller (7.18) is a solution to Problem 2. Therefore, from Lemma 1 there exist positive definite matrices $P_1^0,\ldots,P_{\mathrm{v}}^0$, such that the matrix inequalities (7.20) hold for $\mathscr{B}=\mathscr{B}_0$, $\mathscr{M}=\mathscr{M}_0$, $\mathscr{N}=\mathscr{N}_0$ and $P_i=P_i^0$, $\forall i\in\bar{\mathrm{v}}$. It is to be noted that the quadratic terms with respect to P_i in (7.20) are eliminated, which implies that (7.20) is a LMI with respect to P_i, and thus can be solved using the available LMI solvers.

An algorithm is introduced next, which aims to design a centralized controller solving Problem 2, while it meets the condition (iii) of Problem 1 as well.

Algorithm 1:

Step 1) Set $W=\mathscr{B}_0$ and $V_i=P_i^0$ for all $i\in\bar{\mathrm{v}}$.

Step 2) Minimize the objective function trace(P_1X_0) for the variables $\mathscr{B},\mathscr{M},\mathscr{N}$ and $P_1,\ldots,P_{\mathrm{v}}\geq 0$, subject to the inequality constraints (7.22), which are LMIs (according to Remark 1), and the constraint that $\mathscr{B}$ is block diagonal (note that X_0 is defined in (7.6)).

Step 3) If $\sum_{i=1}^{\nu} \|V_i - P_i\| + \|W - \mathscr{B}\| \leq \delta$, where δ is a prescribed permissible deviation, then stop. Otherwise, set $V_i = P_i$, $i \in \bar{\nu}$, and $W = \mathscr{B}$, and go to Step 2.

Let the matrices $\mathscr{B}$, $\mathscr{M}$ and $\mathscr{N}$ obtained in Algorithm 1 be denoted by $\mathscr{B}_{opt}$, $\mathscr{M}_{opt}$ and $\mathscr{N}_{opt}$, respectively. It can be easily seen that the control (7.18) with the parameters $\mathscr{B}_{opt}$, $\mathscr{M}_{opt}$ and $\mathscr{N}_{opt}$ satisfies the requirements of Problem 2 and the condition (iii) of Problem 1.

Remark 2 *The objective function trace$(P_1 X_0)$ introduced in Step 2 of Algorithm 1 is, in fact, equivalent to the performance index J given by (7.7). The details of this equivalency may be found in [33].*

Remark 3 *As pointed out earlier, the matrix inequalities given by (7.20) are satisfied for $\mathscr{B} = \mathscr{B}_0$, $\mathscr{M} = \mathscr{M}_0$, $\mathscr{N} = \mathscr{N}_0$, and $P_i = P_i^0$, $i \in \bar{\nu}$. On the other hand, by setting $V_i = P_i^0$ and $W = \mathscr{B}_0$, the LMIs (7.22) will be equivalent to the matrix inequalities (7.20). This implies that the LMI problem given in Step 2 of Algorithm 1 is feasible. In addition, it is evident that this algorithm is monotone decreasing and convergent, and should ideally stop when $W = \mathscr{B}$ and $V_i = P_i$ for all $i \in \bar{\nu}$. This results from the conditions under which the inequalities (7.26a) and (7.26b) turn to the equalities. However, Step 3 is required in order for the algorithm to halt in a finite number of iterations.*

The centralized servomechanism controller obtained here will be used in the next section to find a high-performance decentralized servomechanism controller.

7.5 Optimal decentralized servomechanism controller

Consider the centralized controller $\tilde{K}_c$ of the form (7.11) with the following parameters:

$$\begin{aligned} \dot{\eta}_c(t) &= \mathscr{A}\eta_c(t) + \mathscr{B}_{opt}Cx(t) \\ u(t) &= \mathscr{M}_{opt}\eta_c(t) + \mathscr{N}_{opt}x(t) \end{aligned} \tag{7.32}$$

The methodology proposed in Section 7.3 can now be applied to the centralized controller $\tilde{K}_c$ in order to obtain a decentralized controller denoted by $\tilde{K}_d$. For this purpose, let the above controller be decomposed as:

$$\begin{aligned} \dot{\eta}_c(t) &= \mathscr{A}\eta_c(t) + \mathbf{B}_{opt}^{i}\mathbf{C}^{i}x^{i}(t) + \mathbf{B}_{i}^{opt}C_i x_i(t) \\ u^{i}(t) &= \mathbf{M}_{opt}^{i}\eta_c(t) + \mathbf{N}_{opt}^{i}x^{i}(t) + \mathbf{N}_{i}^{opt}x_i(t) \\ u_i(t) &= \mathbf{M}_{i}^{opt}\eta_c(t) + \mathbf{N}_{opt}^{\bar{i}}x^{i}(t) + \mathbf{N}_{\bar{i}}^{opt}x_i(t) \end{aligned} \tag{7.33}$$

where the matrices $\mathbf{C}^{i}, \mathbf{B}_{opt}^{i}, \mathbf{B}_{i}^{opt}, \mathbf{M}_{opt}^{i}, \mathbf{M}_{i}^{opt}, \mathbf{N}_{opt}^{i}, \mathbf{N}_{i}^{opt}, \mathbf{N}_{opt}^{\bar{i}}$ and $\mathbf{N}_{\bar{i}}^{opt}$ are derived from $C, \mathscr{B}_{opt}, \mathscr{M}_{opt}$ and $\mathscr{N}_{opt}$. Therefore, the state-space representation of the local controller $\tilde{K}_{d_i}$, $i \in \bar{\nu}$, will be obtained as follows:

$$\begin{aligned}\dot{\eta}_{d_i}(t) &= \begin{bmatrix} \mathbf{A}^i + \mathbf{B}^i\mathbf{N}^i_{opt} & \mathbf{B}^i\mathbf{M}^i_{opt} \\ \mathbf{B}^i_{opt}\mathbf{C}^i & \mathscr{A} \end{bmatrix} \eta_{d_i}(t) + \begin{bmatrix} \mathbf{A}_i + \mathbf{B}^i\mathbf{N}^{opt}_i \\ \mathbf{B}^{opt}_i C_i \end{bmatrix} x_i(t) \\ u_i(t) &= \begin{bmatrix} \mathbf{N}^{\bar{i}}_{opt} & \mathbf{M}^{opt}_i \end{bmatrix} \eta_{d_i}(t) + \mathbf{N}^{opt}_{\bar{i}} x_i(t)\end{aligned} \tag{7.34}$$

Suppose that the initial state of the controller $\tilde{K}_{d_i}$ is equal to $\eta_{d_i}(0) = \begin{bmatrix} X^{i\,T}_{\mu} & 0_{1\times rp} \end{bmatrix}^T$, for all $i \in \bar{\mathrm{v}}$, where $0_{1\times rp}$ denotes the $1 \times rp$ zero matrix. It is desired to prove that $\tilde{K}_d$ is a solution of Problem 1.

Theorem 4 *The decentralized controller $\tilde{K}_d$ satisfies the requirements (i) and (ii) of Problem 1 for the system $\mathscr{S}$.*

Proof: Since $\tilde{K}_c$ given by (7.32) is designed in Section 7.4 in such a way that it stabilizes the modified system S^i for any $i \in \bar{\mathrm{v}}$, it can be concluded from Theorem 2 that the state $x(t)$ of the system $\mathscr{S}$ under the decentralized controller $\tilde{K}_d$ goes to zero as $t \to \infty$, provided $z(0) = 0$. Thus, the requirement (i) of Problem 1 is met. Denote the block diagonal matrix $\mathscr{B}_{opt}$ as:

$$\mathscr{B}_{opt} = \mathrm{diag}\left(\begin{bmatrix} \mathscr{B}^{opt}_{11} & \mathscr{B}^{opt}_{22} & \cdots & \mathscr{B}^{opt}_{\mathrm{vv}} \end{bmatrix}\right) \tag{7.35}$$

It can be easily verified that $\mathbf{B}^{opt}_i$ introduced in (7.33) is equal to:

$$\mathbf{B}^{opt}_i = \begin{bmatrix} 0_{r_i\times r_1 p} & \cdots & 0_{r_i\times r_{i-1}p} & \mathscr{B}^{opt\,T}_{ii} & 0_{r_i\times r_{i+1}p} & \cdots & 0_{r_i\times r_{\mathrm{v}}p} \end{bmatrix}^T, \quad i \in \bar{\mathrm{v}} \tag{7.36}$$

Furthermore, $\mathbf{B}^i_{opt}\mathbf{C}^i$ is derived from $\mathscr{B}_{opt}C$ by removing its ith block column (which is equal to $\mathbf{B}^{opt}_i C_i$). This observation along with the fact that $\mathscr{B}_{opt}$ and C are block diagonal, yield that the ith block row of $\mathbf{B}^i_{opt}\mathbf{C}^i$ is a zero matrix. Using this result and substituting (7.36) into (7.34), one can rearrange the entries of the state vector $\eta_{d_i}(t)$ in order to come up with the following state-space representation for the local controller $\tilde{K}_{d_i}$:

$$\begin{aligned}\dot{\tilde{\eta}}_{d_i}(t) &= \begin{bmatrix} \mathscr{A}_i & 0 \\ L_{i_1} & L_{i_2} \end{bmatrix} \tilde{\eta}_{d_i}(t) + \begin{bmatrix} \mathscr{B}^{opt}_{ii} & 0 \\ 0 & L_{i_3} \end{bmatrix} \begin{bmatrix} y_i(t) & 0 \\ 0 & x_i(t) \end{bmatrix} \\ u_i(t) &= L_{i_4}\tilde{\eta}_{d_i}(t) + \mathbf{N}^{opt}_{\bar{i}_2} x_i(t)\end{aligned} \tag{7.37}$$

where $\mathscr{A}_i$ is defined in (7.30). Apply now the decentralized controller $\tilde{K}_d$ to the system $\mathscr{S}$. Each interconnection signal coming into the subsystem $\mathscr{S}_i$ from the other subsystems is composed of two main components: one is exponentially decaying (because the requirement (i) of Problem 1 is fulfilled) and hence does not affect the regulation of y_i, and the other one is an unbounded component whose effect is similar to $\omega(t)$ in (7.2). This unbounded component together with the disturbance term $E_i\omega(t)$ can be modeled in the state-space representation of the subsystem $\mathscr{S}_i$ as an embedded term $\tilde{E}_i\omega(t)$, where $\omega(t)$ is obtained from (7.2) with a proper initial condition $z(0)$. As a result, the ith subsystem can be modeled as:

$$\begin{aligned}\dot{x}_i(t) &= A_{ii}x_i(t) + B_iu_i(t) + G_ir_i(t) + \tilde{E}_i\omega(t)\\ y_i(t) &= C_ix_i(t)\end{aligned} \tag{7.38}$$

where $r_i(t)$ represents the exponentially decaying component of the incoming interconnections. Since the structure of $\tilde{K}_{d_i}$ in (7.37) complies with the controller proposed in [31], $y_i(t)$ approaches zero as $t \to \infty$, when the local controller $\tilde{K}_{d_i}$ (given by (7.37)) is applied to the system given by (7.38). This completes the proof. ■

So far, it is shown that the decentralized controller $\tilde{K}_d$ satisfies the requirements (i) and (ii) of Problem 1. The requirement (iii) will be investigated next.

Assume that the centralized controller $\tilde{K}_c$ is applied to the system $\mathscr{S}$. Denote the corresponding performance index (7.7) with J_{opt}. Note that J_{opt} is derived from a constrained optimization problem, and ideally, it is desired to have the same performance for the decentralized control system. However, there is a deviation between the decentralized performance index and J_{opt}. A method will be given next to measure this deviation.

The performance index J associated with the system $\mathscr{S}$ under the decentralized controller $\tilde{K}_d$ can be written as $\mathrm{trace}(P_dX_0^d)$, where P_d is derived from a Lyapunov equation [33], and:

$$X_0^d = E\left\{\begin{bmatrix} x(0)\\ X_\mu^1\\ X_\mu^2\\ \vdots\\ X_\mu^\nu\end{bmatrix}\begin{bmatrix} x(0)^T \ X_\mu^{1^T} \ \dots \ X_\mu^{\nu^T}\end{bmatrix}\right\} = \begin{bmatrix} X_0 & X_\mu X_\mu^{1^T} & \cdots & X_\mu X_\mu^{\nu T}\\ X_\mu^1 X_\mu^T & X_\mu^1 X_\mu^{1^T} & \cdots & X_\mu^1 X_\mu^{\nu T}\\ \vdots & \vdots & \ddots & \vdots\\ X_\mu^\nu X_\mu^T & X_\mu^\nu X_\mu^{1^T} & \cdots & X_\mu^\nu X_\mu^{\nu T}\end{bmatrix} \tag{7.39}$$

According to Theorem 1, if X_μ^i is equal to $x^i(0)$ for all $i \in \bar{\nu}$, then the state and the input of the centralized closed-loop system are the same as those of the corresponding decentralized closed-loop system. Hence, J_{opt} can alternatively be written as $\mathrm{trace}(P_dX_0^c)$, where:

$$X_0^c = E\left\{\begin{bmatrix} x(0)\\ x^1(0)\\ x^2(0)\\ \vdots\\ x^\nu(0)\end{bmatrix}\begin{bmatrix} x(0)^T \ x^1(0)^T \ \dots \ x^\nu(0)^T\end{bmatrix}\right\} \tag{7.40}$$

Therefore, the error between J and J_{opt} can be obtained as follows:

$$J_{opt} - J = \mathrm{trace}\left(P_d\begin{bmatrix} 0 & \mathrm{cov}(X_\mu, X_\mu^1) & \cdots & \mathrm{cov}(X_\mu, X_\mu^\nu)\\ \mathrm{cov}(X_\mu^1, X_\mu) & \mathrm{cov}(X_\mu^1, X_\mu^1) & \cdots & \mathrm{cov}(X_\mu^1, X_\mu^\nu)\\ \vdots & \vdots & \ddots & \vdots\\ \mathrm{cov}(X_\mu^\nu, X_\mu) & \mathrm{cov}(X_\mu^\nu, X_\mu^1) & \cdots & \mathrm{cov}(X_\mu^\nu, X_\mu^\nu)\end{bmatrix}\right) \tag{7.41}$$

where $\text{cov}(\pi_1, \pi_2) = E\left\{\pi_1 \pi_2^T\right\} - E\left\{\pi_1\right\} E\left\{\pi_2^T\right\}$ for any arbitrary column vectors π_1 and π_2. One can use the equations (7.40) and (7.41) to obtain the relative error between J and J_{opt}.

Remark 4 *One can use the equation (7.41) to find out how close the decentralized performance index J is to the optimal centralized counterpart J_{opt}. In addition, it can be deduced from (7.41) that the more the initial state $x(0)$ tends to be deterministic, the closer J becomes to J_{opt}, and in the case of a deterministic initial state, J is equal to J_{opt}. This observation along with the result of Theorem 4 confirm that $\tilde{K}_d$ is a solution of Problem 1.*

Remark 5 *The centralized servomechanism controller $\tilde{K}_c$ is obtained in Section 7.4 from an optimization problem subject to the constraint that all of the systems $\mathbf{S}^2, \mathbf{S}^3, ..., \mathbf{S}^\nu$ are stable under the controller $\tilde{K}_c$. This constraint is essential in the development of the decentralized controller. Nevertheless, when the interconnections between the subsystems of $\mathscr{S}$ are relatively weak, it is expected that the aforementioned constraint may not influence the solution of the optimization problem. In other words, in the presence of sufficiently weak interconnections, $\tilde{K}_c$ obtained by solving the optimization problem automatically stabilizes the systems $\mathbf{S}^2, ..., \mathbf{S}^\nu$. This implies that the above constraint is unlikely to degrade the global optimality of the solution. This point is further clarified in Example 1.*

7.6 Practical considerations in control design

The decentralized servomechanism controller in the previous section was designed based on some rather unrealistic assumptions. For instance, the modeling parameters of the overall system have been assumed to be exactly known in all local subsystems. In order to modify the control design procedure to make it more suitable in a practical framework, the following issues will now be considered in the control design:

- The knowledge of the system parameters is not identical from the viewpoints of different subsystems.
- The system $\mathscr{S}$ is subject to perturbation in the sense that its parameters A and B are uncertain.
- There exist delays in the interconnection and input signals.

A few definitions will now be given in order to develop the remaining results.

Definition 3 *The system $\tilde{\mathbf{S}}^i$, $i \in \bar{\nu}$, is defined to be a system obtained from $\mathscr{S}$ by carrying out the following steps:*

- *All of the interconnections coming into its ith subsystem are removed.*
- *The parameters of the system, except for those of the subsystem $\mathscr{S}_i$ and its outgoing interconnections, are replaced by their corresponding values representing the own belief of the subsystem $\mathscr{S}_i$ about the system $\mathscr{S}$.*

Definition 4 *Let the perturbed model of the system $\mathscr{S}$ be denoted by $\bar{\mathscr{S}}$. The system $\bar{\mathbf{S}}^i$, $i \in \bar{\mathrm{v}}$, is defined to be a system obtained from $\tilde{\mathbf{S}}^i$ by replacing the parameters of its subsystem i together with its own outgoing interconnections, with their corresponding perturbed values of the system $\bar{\mathscr{S}}$. Denote the state-space representation of $\bar{\mathbf{S}}^i$ as:*

$$\begin{aligned} \dot{x}(t) &= \bar{A}^i x(t) + \bar{B}^i u(t) \\ y(t) &= \bar{C}^i x(t) \end{aligned} \tag{7.42}$$

In order to proceed with the control design, all assumptions made earlier for the system $\mathscr{S}$ (Assumptions 1, 2 and 3) are also required to correspondingly hold for the systems $\tilde{\mathbf{S}}^i$, $i \in \bar{\mathrm{v}}$ (because $\tilde{\mathbf{S}}^i$ is the ith subsystem $\mathscr{S}_i$ belief of the system $\mathscr{S}$). Suppose that all of those assumptions are satisfied. For any $i \in \bar{\mathrm{v}}$, design a centralized servomechanism controller $\tilde{K}_c^i$ for the system $\tilde{\mathbf{S}}^i$ using the methodology explained in Section 7.4. Then, convert the centralized controller $\tilde{K}_c^i$ to a decentralized servomechanism controller $\tilde{K}_d^i$ as pointed out in Section 7.5, and denote its local controllers with $\tilde{K}_{d_1}^i, \tilde{K}_{d_2}^i, ..., \tilde{K}_{d_\mathrm{v}}^i$. Define now the decentralized servomechanism controller $\hat{K}_d$ as a controller consisting of the local controllers $\tilde{K}_{d_1}^1, \tilde{K}_{d_2}^2, ..., \tilde{K}_{d_\mathrm{v}}^\mathrm{v}$. It is to be noted that $\hat{K}_d$ is a modified version of the controller $\tilde{K}_d$ obtained in Section 7.5, as it is attained through a procedure which takes into account the practical issues such as perturbations and non identical beliefs of the subsystems about the system parameters. It is interesting to observe that in the case when any two different subsystems think of the system $\mathscr{S}$ identically, $\tilde{K}_{d_l}^i$ is the same as $\tilde{K}_{d_l}^j$, for any $i, j, l \in \bar{\mathrm{v}}$.

Definition 5 *Let $\hat{\mathscr{S}}$ represent the system obtained from $\bar{\mathscr{S}}$ by considering the delay d_{i_j} for the jth input of its ith subsystem, for any $i \in \bar{\mathrm{v}}$ and $j \in \{1, 2, ..., m_i\}$. Accordingly, $\hat{\mathbf{S}}^i$, $i \in \bar{\mathrm{v}}$, is defined to be a system obtained from $\bar{\mathbf{S}}^i$ by considering the delay d_{i_j} for the jth input of its ith subsystem, for any $j \in \{1, 2, ..., m_i\}$.*

It is to be noted that $\hat{\mathscr{S}}$ is the delayed and perturbed version of the ideal system $\mathscr{S}$. Moreover, $\hat{\mathbf{S}}^i$, $i \in \bar{\mathrm{v}}$, is derived from the system $\tilde{\mathbf{S}}^i$ by imposing the delays and perturbations on its ith subsystem only. It is also worth noting that $\hat{\mathscr{S}}$ is, in fact, a more accurate model for the system represented by $\mathscr{S}$, in a practical environment. An important characteristic of this system will be investigated in the next theorem.

Theorem 5 *The system $\hat{\mathscr{S}}$ is stable under the decentralized controller $\hat{K}_d$ if and only if the system $\hat{\mathbf{S}}^i$ is stable under the centralized controller $\tilde{K}_c^i$ for any $i \in \bar{\mathrm{v}}$.*

Proof: The proof is omitted due to its similarity to the proof of Theorem 2. ■

Remark 6 *Theorem 5 translates the stability of the decentralized control system into that of a set of centralized control systems. This alternative approach is useful to obtain some permissible bounds on the uncertain parameters of the system $\hat{\mathscr{S}}$ and the delays. For instance, in case of delay-free systems, the problem can be reduced to finding the sensitivity of the eigenvalues of a number of matrices with respect to the variations in their entries. This has been addressed in the literature using* *different mathematical approaches [34], [35], and it is known that this sensitivity depends on several factors such as the norm of the perturbation matrix, structure of the matrix (represented by condition number or eigenvalue condition number [34]), and repetition or distinction of the eigenvalues, in general.*

Theorem 6 *Assume that the system $\hat{\mathscr{S}}$ is stable under the decentralized controller $\hat{K}_d$. The output $y(t)$ is regulated to zero for any $z(0)$.*

Proof: The proof is similar to the one given in [20] for the asymptotic regulation property of the delay-free systems, and is omitted here. ■

Theorem 6 states that as long as the system $\hat{\mathscr{S}}$ is stable under the decentralized servomechanism controller $\hat{K}_d$, the desired output regulation is achieved. Hence, the sole concern regarding the controller $\hat{K}_d$ is that it should maintain the stability of the closed-loop system, which can be ensured by obtaining a number of conditions using Remark 6 and the references therein.

Remark 7 *The deviation in the performance index for the case when all individual subsystems assume the same modeling parameters for the system $\mathscr{S}$ was obtained in Section 7.5, and is expressed by the equation (7.41). However, one can pursue the same methodology in order to attain a similar result under the assumptions made in this section, which describe a more pragmatic case.*

7.7 Numerical examples

Example 1: Consider a system $\mathscr{S}$ consisting of two interconnected subsystems with the following state-space representation for its first subsystem S_1:

$$\begin{aligned} \dot{x}_1(t) &= \begin{bmatrix} 1 & -2 \\ 2 & 3 \end{bmatrix} x_1(t) + \begin{bmatrix} 1 \\ 3 \end{bmatrix} u_1(t) + E_1\omega(t), \\ y_1(t) &= \begin{bmatrix} -1 & 2 \end{bmatrix} x_1(t) \end{aligned} \tag{7.43}$$

and the following representation for its second subsystem S_2:

$$\begin{aligned} \dot{x}_2(t) &= \begin{bmatrix} -1 & 2 \end{bmatrix} x_1(t) - 3x_2(t) + 5u_2(t) + E_2\omega(t), \\ y_2(t) &= 3x_2(t) \end{aligned} \tag{7.44}$$

where :

- $\omega(t)$ is assumed to be the scalar exponential function e^t, which represents the structure of the disturbance affecting the input of the system.
- E_1 and E_2 are unknown matrices of proper dimensions, which account for the unmeasurable nature of the disturbance in the system.

Assume that the initial state of the system is a random variable with X_0 (defined in (7.6)) equal to I. It is desired to design a decentralized controller K_d to solve Problem 1 under the assumption $Q = R = I$. To this end, an initial centralized controller which can reject the disturbance $\omega(t)$ is to be designed first. This controller is obtained (using the method proposed earlier) with the parameters given below:

$$\mathscr{A} = \begin{bmatrix} 1 & 0 \\ 0 & 1 \end{bmatrix}, \quad \mathscr{B}_o = \begin{bmatrix} 1 & 0 \\ 0 & 1 \end{bmatrix},$$
$$\mathscr{M}_o = \begin{bmatrix} -3.3182 & 0 \\ 0 & -1.0393 \end{bmatrix}, \quad \mathscr{N}_o = \begin{bmatrix} 0.9231 & -4.4856 & 0 \\ 0 & 0 & -1.0147 \end{bmatrix} \tag{7.45}$$

Using Algorithm 1 for optimizing the performance of the initial controller, one will arrive at a centralized controller $\tilde{K}_c$ described in (7.32) with the state-space matrices:

$$\mathscr{B}_{opt} = \begin{bmatrix} 3.3253 & 0 \\ 0 & 1.6471 \end{bmatrix}, \quad \mathscr{M}_{opt} = \begin{bmatrix} -0.9348 & -0.0207 \\ 0.0988 & -0.5580 \end{bmatrix},$$
$$\mathscr{N}_{opt} = \begin{bmatrix} 0.8214 & -4.2823 & -0.0513 \\ 0.0480 & -0.1015 & -0.9764 \end{bmatrix} \tag{7.46}$$

The resultant quadratic performance index J corresponding to the initial controller given to Algorithm 1 and the optimal controller $\tilde{K}_c$ are given by 8.6425 and 3.9422, respectively. This sizable reduction in the cost function points to the effectiveness of Algorithm 1. Now, let decentralize the controller $\tilde{K}_c$ using the procedure proposed in Section 7.5, to obtain the local controllers $\tilde{K}_{d_1}$ and $\tilde{K}_{d_2}$ described by:

$$\dot{\eta}_{d_1}(t) = \begin{bmatrix} -7.8820 & 0.4940 & -2.7898 \\ 0 & 1.0000 & 0 \\ 4.9412 & 0 & 1.0000 \end{bmatrix} \eta_{d_1}(t) + \begin{bmatrix} -0.7602 & 1.4925 \\ -3.3253 & 6.6506 \\ 0 & 0 \end{bmatrix} x_1(t)$$
$$u_1(t) = \begin{bmatrix} -0.0513 & -0.9348 & -0.0207 \end{bmatrix} \eta_{d_1}(t) + \begin{bmatrix} 0.8214 & -4.2823 \end{bmatrix} x_1(t) \tag{7.47}$$

and:

$$\dot{\eta}_{d_2}(t) = \begin{bmatrix} 1.8214 & -6.2823 & -0.9348 & -0.0207 \\ 4.4642 & -9.8468 & -2.8043 & -0.0621 \\ -3.3253 & 6.6506 & 1.0000 & 0 \\ 0 & 0 & 0 & 1.0000 \end{bmatrix} \eta_{d_2}(t) + \begin{bmatrix} -0.0513 \\ -0.1539 \\ 0 \\ 4.9412 \end{bmatrix} x_2(t)$$
$$u_2(t) = \begin{bmatrix} 0.0480 & -0.1015 & 0.0988 & -0.5580 \end{bmatrix} \eta_{d_2}(t) - 0.9764 x_2(t) \tag{7.48}$$

respectively. It is worth mentioning that these local controllers are attained based upon the assumption that every subsystem knows the parameters of the other subsystem, but not necessarily its initial state. To evaluate the performance of the controller $\tilde{K}_d$, suppose that the real initial state $x(0)$ is equal to $\begin{bmatrix} 1.5 & 1.5 & 1.5 \end{bmatrix}^T$. This represents an inferior scenario in light of the relation $X_0 = I$ (in fact, it can be easily verified that the initial state given above is noticeably far from its mean). Now, consider two cases as follows:

- Assume that:

$$E_1 = \begin{bmatrix} 1 \\ 3 \end{bmatrix}, \quad E_2 = \begin{bmatrix} 5 \end{bmatrix} \tag{7.49}$$

and that each local controller knows the initial state of the other subsystem with -100% error. As a result, the initial states η_{d_1} and η_{d_2} are zero vectors. Let the external input $\sin(3t)$ be applied to the system $\mathscr{S}$, other than the input disturbance. The outputs of the first and the second subsystems of the system $\mathscr{S}$ under the controllers $\tilde{K}_c$ and $\tilde{K}_d$ are depicted in Figure 7.1. As it can be observed, these two controllers perform almost identically such that the discrepancy in their corresponding signals is barely visible (specially in the output $y_1(t)$). This figure also illustrates that the disturbance is rejected very quickly and that the steady-state trajectory is reached rather shortly, although the prediction error was significantly large.

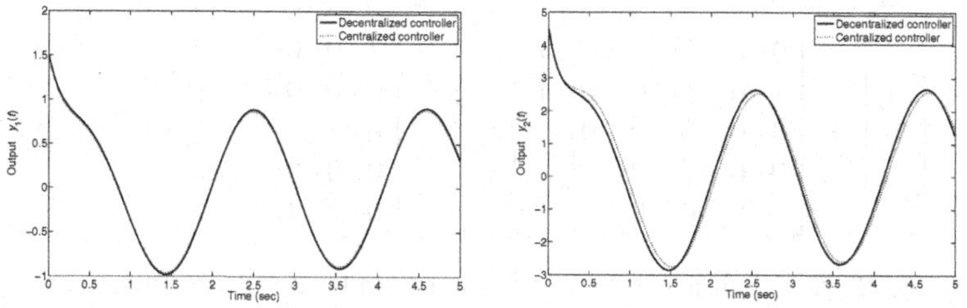

Fig. 7.1 The outputs $y_1(t)$ and $y_2(t)$ of the centralized and decentralized control systems in the presence of -100% prediction error for the initial state.

- Assume that E_1 and E_2 are the same as the ones given in (7.49), and that each local controller knows the initial state of the other subsystem with 5000% error (i.e. an extremely severe situation is considered here). Hence,

$$\eta_{d_1} = \begin{bmatrix} 75\ 0\ 0 \end{bmatrix}, \quad \eta_{d_2} = \begin{bmatrix} 75\ 75\ 0\ 0 \end{bmatrix} \tag{7.50}$$

  Let the external unbounded input $t \times \sin(t)$ be applied to the system $\mathscr{S}$. The outputs of the first and the second subsystems of the system $\mathscr{S}$ under the controllers $\tilde{K}_c$ and $\tilde{K}_d$ are depicted in Figure 7.2 to substantiate how insensitive the decentralized controller $\tilde{K}_d$ to prediction error is.

Example 2: Consider a leader-follower formation control system consisting of three vehicles. Assume that the state of each vehicle is available for its follower vehicle in order to generate the corresponding control signal. The objective is to design a controller such that all of the vehicles fly at the same desired speed, while the desired Euclidean distances between vehicles are achieved. The exact linearized model for the aforementioned problem (using certain specifications for each vehicle) is obtained in [4], and the tracking problem is converted to a regulation problem represented by a system $\mathscr{S}$ with the following state-space equation:

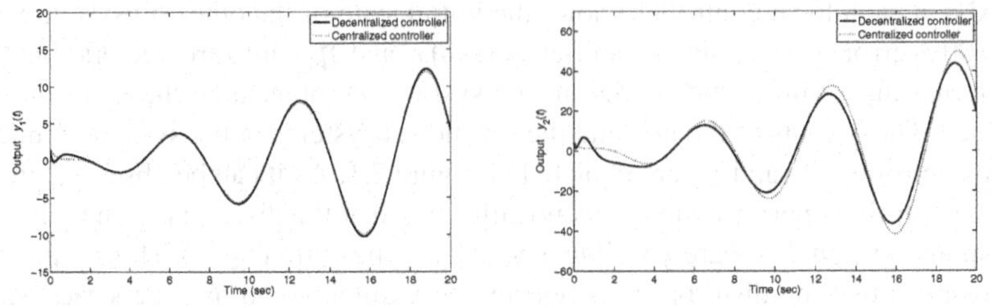

Fig. 7.2 The outputs $y_1(t)$ and $y_2(t)$ of the centralized and decentralized control systems in the presence of 5000% prediction error for the initial state.

$$\begin{bmatrix} \dot{x}_1 \\ \dot{x}_2 \\ \dot{x}_3 \end{bmatrix} = \begin{bmatrix} 0_2 & 0_2 & 0_2 & 0_2 & 0_2 \\ I_2 & 0_2 & -I_2 & 0_2 & 0_2 \\ 0_2 & 0_2 & 0_2 & 0_2 & 0_2 \\ 0_2 & 0_2 & I_2 & 0_2 & -I_2 \\ 0_2 & 0_2 & 0_2 & 0_2 & 0_2 \end{bmatrix} \begin{bmatrix} x_1 \\ x_2 \\ x_3 \end{bmatrix} + \begin{bmatrix} I_2 & 0_2 & 0_2 \\ 0_2 & 0_2 & 0_2 \\ 0_2 & I_2 & 0_2 \\ 0_2 & 0_2 & 0_2 \\ 0_2 & 0_2 & I_2 \end{bmatrix} \begin{bmatrix} u_1 \\ u_2 \\ u_3 \end{bmatrix} \tag{7.51}$$

where I_2 and 0_2 represent a 2×2 identity matrix and a 2×2 zero matrix, respectively, and

$$x_1 = \begin{bmatrix} x_{11} \\ x_{12} \end{bmatrix}, \ x_2 = \begin{bmatrix} x_{21} \\ x_{22} \\ x_{23} \\ x_{24} \end{bmatrix}, \ x_3 = \begin{bmatrix} x_{31} \\ x_{32} \\ x_{33} \\ x_{34} \end{bmatrix} \tag{7.52}$$

and where $u_i = \begin{bmatrix} u_{i1} \\ u_{i2} \end{bmatrix}$, $i = 1, 2, 3$. Note that the system $\mathscr{S}$ consists of three subsystems: one leader and two follower vehicles. x_1 denotes the state of the leader, and x_2 and x_3 represent the state of vehicles 2 and 3 (i.e., the followers), respectively. More specifically:

1. x_{11} and x_{12} are the speed error of the leader (speed of the leader minus its desired speed) along x and y axes, respectively.
2. x_{i1} and x_{i2}, $i = 2, 3$, are the distance error (distance between vehicles i and $i-1$ minus the corresponding desired distance) along x and y axes, respectively.
3. x_{i3} and x_{i4}, $i = 2, 3$, are the speed error (speed of vehicle i minus its desired speed) along x and y axes, respectively.
4. u_{i1} and u_{i2}, $i = 1, 2, 3$, are the acceleration of vehicle i along x and y axes, respectively.

Note that the system is assumed to be disturbance free. It is desired now to design a decentralized controller for $\mathscr{S}$, such that the closed-loop system is stable. Moreover, the objective is that the state variables of the closed-loop system decay as sharply as possible, with a reasonably small control effort. To attain these specifications, consider the performance index given by (7.7), and assume that $Q = R = I$. Two different design techniques will be used and the results will be compared here:

the iterative numerical procedure given in [17], and the method proposed here. Suppose that each initial state is uniformly distributed in the intervals [200, 400], and that any two distinct initial state variables are statistically independent. It is to be noted that the units used for distance and velocity in the state vectors are ft and ft/s, respectively. Assume that any two different subsystems consider the same expected value for the initial state of the remaining subsystem, and that the model of each subsystem is exactly known by the other subsystems. Design an optimal centralized controller for the system $\mathscr{S}$ by using the Riccati equation, and denote it with K_c. Note that this controller is independent of the initial state. It is straightforward to show that the modified systems $\mathbf{S}^2$ and $\mathbf{S}^3$ are both stable under the controller K_c. This result can be intuitively deduced due to the weak interconnections between subsystems, as pointed out in Remark 5. Now, employing the method given in Section 7.3, one can design a decentralized controller K_d in terms of K_c.

To compare the centralized controller K_c, the decentralized controller K_d, and the decentralized controller obtained from [17] with each other, assume that the real initial state variables are all equal to 400, which correspond, in fact, to the worst case scenario (maximum discrepancy between the real initial state variables, i.e. 400, and the corresponding expected values, i.e. 300, which are used by the proposed controller). Consider the following performance index as a benchmark:

$$J^* = \int_0^\infty \left(x(t)^T Q x(t) + u(t)^T R u(t)\right) dt \tag{7.53}$$

Note that unlike the (stochastic) LQ performance index J in (7.7), J^* does not include the operator $E\{\cdot\}$. The iterative numerical procedure of [17] gives a static decentralized state feedback law which results in $J^* = 2{,}257{,}085$. On the other hand, J^* obtained by applying the method proposed in this chapter is equal to 2,090,939, while the best achievable performance index J^* corresponding to the centralized LQR controller is equal to 2,068,513. This means that the relative errors of the performance indices obtained by using the methods given here and the one in [17], with respect to the optimal centralized performance index, are 1.08% and 9.12%, respectively. This shows clearly that the controller proposed in this chapter outperforms the one presented in [17], significantly.

Figures 7.3 depict the time responses of the system under the controller proposed in this chapter (dotted curve), the controller proposed in [17] (dashed curve), and the optimal centralized controller (solid curve) for three state variables x_{11}, x_{31}, x_{33}. Moreover, the control signals u_{11}, u_{21}, u_{31} obtained by using the three methods discussed above are depicted in Figures 7.4 in a similar way. It is to be noted that despite the relatively big differences between the real initial variables (400 ft for distance errors and 400 ft/sec for speed errors) and the corresponding expected values (300 ft for distance errors and 300 ft/sec for speed errors), which are used to construct the proposed controller, the results obtained by utilizing the proposed method are reasonably close to the time response of the system under the LQR controller.

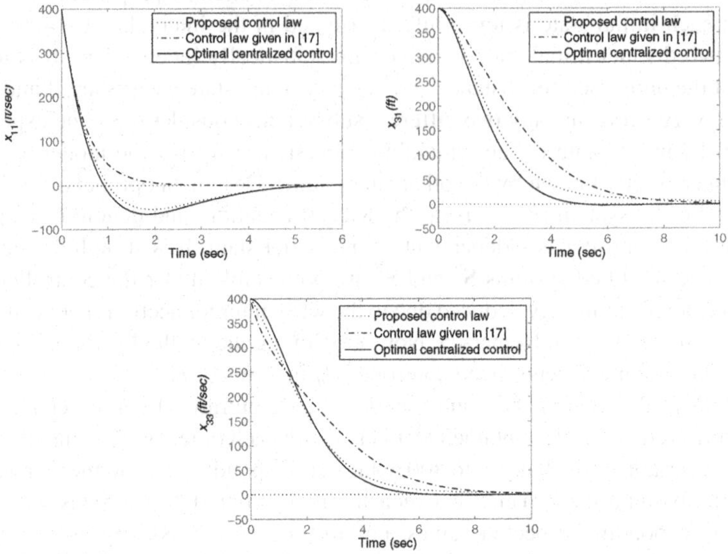

Fig. 7.3 The state variables x_{11}, x_{31} and x_{33} obtained by using three different design techniques in Example 2.

7.8 Summary

In this chapter, a near-optimal decentralized servomechanism controller is designed for a hierarchical interconnected system. This controller results in a satisfactory performance with respect to a predetermined LQ cost function, and is capable of rejecting unmeasurable external disturbances of known dynamics. The case when the system is subject to perturbation and input delays is also investigated, and necessary and sufficient conditions to achieve the stability and disturbance rejection for the closed-loop system are obtained. The designed controller relies on the information of every individual subsystem about the overall system, and since this information is inexact in practice, a procedure is presented to assess the degradation of the performance of the decentralized control system as a result of the erroneous information. The simulation results demonstrate the effectiveness of the present work.

References

1. S. Sojoudi, J. Lavaei and A. G. Aghdam, "Optimal information flow structure for control of interconnected systems," in *Proceedings of 2007 American Control Conference*, New York, NY, 2007.

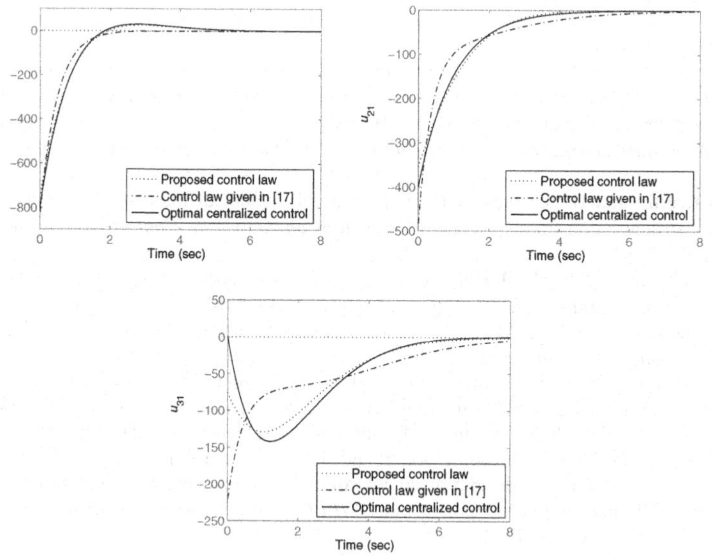

Fig. 7.4 The control signals u_{11}, u_{21} and u_{31} obtained by using three different design techniques in Example 2.

2. A. G. Aghdam, E. J. Davison and R. B. Arreola, "Structural modification of systems using discretization and generalized sampled-data hold functions," *Automatica*, vol. 42, no. 11, pp. 1935-1941, 2006.
3. G. Inalhan, D. M. Stipanovic and C. J. Tomlin, "Decentralized optimization with application to multiple aircraft coordination," in *Proceedings of 41st IEEE Conference on Decision and Control*, Las Vegas, NV, 2002.
4. D. M. Stipanovic, G. Inalhan, R. Teo and C. J. Tomlin, "Decentralized overlapping control of a formation of unmanned aerial vehicles," *Automatica*, vol. 40 , no.8, pp. 1285-1296, 2004.
5. S. S. Stankovic, M. J. Stanojevic and D. D. Šiljak, "Decentralized overlapping control of a platoon of vehicles," *IEEE Transactions on Control Systems Technology*, vol. 8, no. 5, pp. 816-832, 2000.
6. J. A. Fax and R. M. Murray, "Information flow and cooperative control of vehicle formations," *IEEE Transactions on Automatic Control*, vol. 49, no. 9, pp. 1465-1476, 2004.
7. J. Lavaei and A. G. Aghdam, "A necessary and sufficient condition for the existence of a LTI stabilizing decentralized overlapping controller," in *Proceedings of 45th IEEE Conference on Decision and Control*, San Diego, CA, 2006.
8. J. Lavaei and A. G. Aghdam, "High-performance decentralized control design for general interconnected systems with applications in cooperative control," *International Journal of Control*, vol. 80, no. 6, pp. 935-951, 2007.
9. D. D. Šiljak, Decentralized control of complex systems, Boston: Academic Press, 1991.
10. E. J. Davison and T. N. Chang, "Decentralized stabilization and pole assignment for general proper systems," *IEEE Transactions on Automatic Control*, vol. 35, no. 6, pp. 652-664, 1990.
11. J. Lavaei, A. Momeni and A. G. Aghdam, "A model predictive decentralized control scheme with reduced communication requirement for spacecraft formation, *IEEE Transactions on Control Systems Technology*, vol. 16, no. 2, pp. 268-278, 2008.
12. Z. Gong and M. Aldeen, "Stabilization of decentralized control systems," *Journal of Mathematical Systems, Estimation, and Control*, vol. 7, no. 1, pp. 1-16, 1997.

13. J. Lavaei and A. G. Aghdam, "Elimination of fixed modes by means of high-performance constrained periodic control," in *Proceedings of 45th IEEE Conference on Decision and Control*, San Diego, CA, 2006.
14. J. Lavaei and A. G. Aghdam, "Characterization of decentralized and quotient fixed modes via graph theory," *Proceedings of 2007 American Control Conference*, New York, NY, 2007.
15. J. Leventides and N. Karcanias, "Decentralized dynamic pole assignment with low-order compensators," *IMA Journal of Mathematical Control and Information*, vol. 24, no. 3, pp. 395-410, 2007.
16. S. S. Keerthi and H. S. Phatak, "Regional pole placement of multivariable systems under control structure constraints," *IEEE Transactions on Automatic Control*, vol. 40, no. 2, pp. 272-276, 1995.
17. H. T. Toivoneh and P. M. Makila, "A descent anderson-moore algorithm for optimal decentralized control," *Automatica*, vol. 21, no. 6, pp. 743-744, 1985.
18. M. Rotkowitz and S. Lall, "A characterization of convex problems in decentralized Control," *IEEE Transactions on Automatic Control*, vol. 51, no. 2, pp. 274-286, 2006.
19. J. Lavaei and A. Aghdam, "Simultaneous LQ control of a set of LTI systems using constrained generalized sampled-data hold functions," *Automatica*, vol. 43, no. 2, pp. 274-280, 2007.
20. E. J. Davison, "The robust decentralized control of a general servomechanism problem," *IEEE Transactions on Automatic Control*, vol. 21, no. 1, pp. 14-24, 1976.
21. E. J. Davison, "The robust decentralized control of a servomechanism problem for composite systems with input-output interconnections," *IEEE Transactions on Automatic Control*, vol. 24, no. 2, pp. 325-327, 1979.
22. E. J. Davison, and Ü. Özgüner, "Synthesis of the decentralized robust servomechanism problem using local models," *IEEE Transactions on Automatic Control*, vol. 27, no. 3, pp. 583-600, 1982.
23. S. V. Savastuk and D. D. Šiljak, "Optimal decentralized control," in *Proceedings of 1994 American Control Conference*, Baltimore, MD, 1994.
24. D. D. Sourlas and V. Manousiouthakis, "Best achievable decentralized performance," *IEEE Transactions on Automatic Control*, vol. 40, no. 11, pp. 1858-1871, 1995.
25. R. Krtolica and D. D. Šiljak, "Suboptimality of decentralized stochastic control and estimation," *IEEE Transactions on Automatic Control*, vol. 25, no. 1, pp. 76-83, 1980.
26. J. R. Broussard, "An approach to the optimal output feedback initial stabilizing gain problem," in *Proceedings of 29th IEEE Conference on Decision and Control*, Honolulu, HI, 1990.
27. P. M. Makila and H. T. Toivoneh, "Computational methods for parametric LQ problems- a survey," *IEEE Transactions on Automatic Control*, vol. 32, no. 8, pp. 658-671, 1987.
28. A. İftar and Ü. Özgüner, "An optimal control approach to the decentralized robust servomechanism problem," *IEEE Transactions on Automatic Control*, vol. 34, no. 12, pp. 1268-1271, 1989.
29. R. S. Smith and F. Y. Hadaegh, "Control topologies for deep space formation flying spacecraft," in *Proceedings of 2002 American Control Conference*, Anchorage, AK, pp. 2836-2841, 2002.
30. K. P. Groves, D. O. Sigthorsson, A. Serrani and S. Yurkovich , "Reference command tracking for a linearized model of an air-breathing hypersonic vehicle," in *AIAA Guidance, Navigation, and Control Conference and Exhibit*, San Francisco, CA, 2005.
31. E. J. Davison, "A generalization of the output control of linear multivariable systems with unmeasurable arbitrary disturbances," *IEEE Transactions on Automatic Control*, vol. 20, no. 6, pp. 788-792, 1975.
32. J. Lam and Y. Y. Cao, "Simultaneous linear-quadratic optimal control design via static output feedback," *International Journal of Robust and Nonlinear Control*, vol. 9, pp. 551-558, 1999.
33. H. Kwakernaak and R. Sivan, Linear optimal control systems, John Wiley & sons, 1972.
34. S. C. Eisenstat and I. C. F. Ipsen, "Three absolute perturbation bounds for matrix eigenvalues imply relative bounds," *SIAM Journal on Matrix Analysis and Applications*, vol. 20, no. 1, pp. 149-158, 1998.
35. Jos. L. M. Van Dorsselaer, "Several concepts to investigate strongly nonnormal eigenvalue problems," *SIAM Journal on Scientific Computing*, vol. 24, no. 3, pp. 1031-1053, 2003.

Chapter 8
Decentralized Implementation of Centralized Controllers for Interconnected Systems

8.1 Introduction

Many real-world systems such as communication networks, large-space structures, power systems, and chemical processes can be modeled as interconnected systems with homogeneous or heterogeneous interacting subsystems [1, 2, 3, 4, 5]. The classical control techniques often fail to control such systems, in light of some well-known computation or communication constraints. This has given rise to the emergence of the decentralized control area that aims to design non-classical structurally constrained controllers [6]. A decentralized controller comprises a set of non-interacting local controllers corresponding to disparate subsystems. The analysis and synthesis of a decentralized control system has long been studied by many researchers. In particular, the decentralized control theory has been recently developed for systems with geographically distributed subsystems in the context of distributed control for diverse applications, such as flight formation [7], consensus [8, 9] and Internet congestion control [10].

To study the decentralized stabilization problem, the notion of decentralized fixed modes (DFM) was introduced in [11] to characterize those modes of a system that cannot be moved using a linear time-invariant (LTI) decentralized controller. Several methods have been proposed in the literature to find the DFMs of a system [12, 13, 14]. The notion of a quotient fixed modes (QFM) was subsequently introduced in [15] to investigate the stabilizability of a system with respect to all (nonlinear and time-varying) decentralized controllers.

Although there has been a plethora of research on finding the best achievable decentralized performance, several related problems are still open or partially solved [6, 16, 17, 18, 19]. The main reason is that while many control problems, such as H_2 or H_∞ optimal controller design, have explicit solutions in the centralized case, they are cumbersome and generally nonconvex when restricted to the class of decentralized controllers. The work [20] provides a lower bound on the achievable decentralized H_2 performance for stable discrete-time linear systems with stable finite zeros. The problem of designing a decentralized controller that achieves certain H_∞ requirements on all subsystems as well as the overall system is tackled in [21], where sufficient conditions are derived. The papers [22] and [23] obtain sufficient nonlinear conditions for the existence of a stabilizing decentralized controller with a guaranteed H_∞ performance (the sensors and actuators are also allowed to fail in [22]). To find the best achievable decentralized H_∞ performance, an infinite-dimensional optimization problem is proposed in [24] based on the parameterization of all decentralized stabilizing controllers, and it is then truncated to a nonconvex finite-dimensional optimization problem. The existence of a decentralized controller providing certain closed-loop properties for a stable system is studied in [25]. A closely related decentralized control problem is also tackled in [26], where time-domain performance limitations are obtained for open-loop stable, square, linear systems, which can be used to bound the settling time and undershoot in decentralized architecture control schemes. Sufficient conditions are derived in [18] to make the decentralized H_2 optimal control problem convex.

The above-mentioned decentralized control problems can be asked in a broader, unified context as follows: given a centralized controller associated with a system, what is the best decentralized controller that generates state and input trajectories for the system that are sufficiently close to those generated by the prescribed centralized controller? This question is partially answered in the literature. The paper [27] proposes a technique to design a static decentralized controller in terms of a prescribed centralized one, but the centralized and decentralized closed-loop performances can be very different. The work [28] aims to design a decentralized controller based on a given centralized controller such that the associated sensitivity functions are close to each other. In order to make the problem convex, that work minimizes a weighted H_2 error whose weighting factor is forced to be dependent on the parameters of the unknown decentralized controller, which leads to an undesirable performance measure. The paper [29] derives sufficient conditions for the approximation of a static centralized controller by a decentralized controller, which requires solving a set of nonlinear algebraic equations. The recent work [19] proposes a method to implement a given centralized controller in a decentralized fashion, which is successfully applied to the flight formation problem in [4]. However, it is not guaranteed that the system behaves similarly under both the given centralized controller and its decentralized counterpart. The primary objective of the current chapter is to address the aforementioned decentralization question for strongly connected LTI systems, i.e. those LTI systems whose subsystems cannot be renumbered in such a way that the corresponding transfer function matrix becomes upper/lower block triangular.

Given a centralized controller for a strongly connected LTI system, the objective of this chapter is to study the existence of a decentralized controller such that the input and state trajectories of the system under this decentralized controller are arbitrarily close to those of the system under the given centralized controller. To this end, it is shown that under mild conditions, there exists such a decentralized controller composed of high-level and low-level decentralized sub-controllers. The control law of the high-level sub-controller is given explicitly, but the low-level sub-controller is designed based on a new notion of *structural initial value observability*. The developed method is then applied to the optimal LQR decentralized control problem through an example. The problem studied in this work encompasses the ones investigated in [4, 19, 27, 29], and is also related to different problems surveyed earlier on achievable decentralized performance. However, unlike the aforementioned works, the present chapter does not derive nonconvex or conservative sufficient conditions. Instead, it shows that every generic centralized controller can be approximated arbitrarily well by a two-level decentralized controller, provided the system satisfies certain mild (easy-to-check) conditions.

8.2 Problem formulation

Consider an interconnected system $\mathscr{S}$ composed of ν interacting LTI subsystems $S_1, S_2, ..., S_\nu$. Let the system $\mathscr{S}$ be governed by the differential equation

$$\begin{aligned} \dot{x}(t) &= Ax(t) + \sum_{i=1}^{\nu} B_i u_i(t), \\ y_j(t) &= C_j x(t), \quad \forall j \in \mathbf{v} := \{1,2,...,\nu\}, \end{aligned} \tag{8.1}$$

where $x(t) \in \mathbf{R}^n$ represents the state of the system $\mathscr{S}$, and $u_j(t) \in \mathbf{R}^{m_j}$ and $y_j(t) \in \mathbf{R}^{r_j}$ are the input and output of the subsystem S_j, respectively, for every $j \in \mathbf{v}$. Denote the initial state $x(0)$ with x_0 and assume that x_0 is a random variable. Define

$$\begin{aligned} B &:= \begin{bmatrix} B_1 \ B_2 \ \cdots \ B_\nu \end{bmatrix}, \quad C := \begin{bmatrix} C_1^T \ C_2^T \ \cdots \ C_\nu^T \end{bmatrix}^T, \\ m &:= m_1 + m_2 + \cdots + m_\nu, \quad r := r_1 + r_2 + \cdots + r_\nu. \end{aligned} \tag{8.2}$$

Note that although the matrices B and C are block-diagonal for many applications especially when each subsystem has its own sub-state (such as in flight formation), these matrices are considered to be general and unconstrained in the present work. Consider a given stabilizing centralized controller K_c of order n_o with the control law

$$\begin{aligned} \dot{z}(t) &= A_o z(t) + B_o y(t), \\ u(t) &= C_o z(t) + D_o y(t), \end{aligned} \tag{8.3}$$

where $z(0) = 0$. In the centralized closed-loop system obtained by applying the controller K_c to the system $\mathscr{S}$, represent different signals as follows:

- Let $x_c(t)$, $u_c(t)$ and $y_c(t)$ denote the state, input and output of the system $\mathscr{S}$, respectively.
- Let $z_c(t)$ denote the state of the controller K_c.
- Let $u_{c_i}(t)$ and $y_{c_i}(t)$ denote the input and output of the subsystem S_i, respectively, for every $i \in \mathbf{v}$.

It is desired to investigate whether there exists a decentralized controller K_d such that the system $\mathscr{S}$ under K_d generates state and input trajectories sufficiently close to the centralized trajectories $x_c(t)$ and $u_c(t)$, respectively. To this end, a new notion will be introduced in the sequel.

Definition 1 *The controller K_c is said to be decentrally implementable if there exists a natural number μ and a stabilizing linear decentralized controller $K_d(\xi)$ parameterized in terms of a multivariate parameter $\xi \in \mathbf{R}^{\mu}$ such that for every given positive reals ε and Δ, there exists a vector $\xi_0 \in \mathbf{R}^{\mu}$ for which the relation*

$$\int_{\varepsilon}^{\infty} \left(\|x_c(t) - x_d(t)\|^2 + \|u_c(t) - u_d(t)\|^2\right) dt < \Delta \tag{8.4}$$

holds if $x_d(t)$ and $u_d(t)$ denote the state and the input of the system $\mathscr{S}$ under the controller $K_d(\xi_0)$, respectively, and $\|\cdot\|$ represents an arbitrary vector norm.

Regarding the parameterized controller $K_d(\xi)$ in Definition 1, a switching-type nonlinear controller $K_d(\xi)$ can be designed using the technique proposed in [30] to make the inequality (8.4) hold; however, this definition requires $K_d(\xi)$ to be linear due to the linearity of the original controller K_c. It is noteworthy that the notion of *decentralized implementation* is instrumental in understating the gap between the achievable centralized and decentralized performances.

The objective of this chapter is twofold. First, it is desired to prove that the given controller K_c is decentrally implementable under mild conditions. Second, it is aimed to construct a parameterized controller $K_d(\xi)$ associated with K_c.

8.3 Main results

Denote the modes of the system $\mathscr{S}$ with $\lambda_1, \lambda_2, ..., \lambda_n$. Define the structural graph of the system $\mathscr{S}$ to be a directed graph with ν vertices such that for every $i, j \in \nu$, $i \neq j$, vertex i is connected to vertex j by a directed edge if the transfer function $C_j(sI - A)^{-1}B_i$ is nonzero. The system $\mathscr{S}$ is said to be *strongly connected* if its associated structural graph is strongly connected, meaning that there exist directed paths from every vertex to all remaining vertices of the graph [15]. A few technical assumptions are required for the development of this chapter, as provided below.

Assumption 1 *The system $\mathscr{S}$ has no decentralized fixed mode, i.e., it is controllable, observable and the inequality*

$$rank \begin{bmatrix} A - \lambda_i I & B_{j_1} & \dots & B_{j_p} \\ C_{j_{p+1}} & 0 & \dots & 0 \\ \vdots & \vdots & \ddots & \vdots \\ C_{j_\nu} & 0 & \dots & 0 \end{bmatrix} \geq n, \qquad \forall i \in \mathbf{n} := \{1, 2, ..., n\} \tag{8.5}$$

holds for every permutation $(j_1, j_2, ..., j_\nu)$ of the set $\{1, 2, ..., \nu\}$ and $p \in \{1, ..., \nu - 1\}$ [13].

Assumption 2 *The system $\mathscr{S}$ is strongly connected.*

Assumption 3 *The inequality $m_j \leq r$ holds for all $j \in \mathbf{v}$ and, in addition, there exists a matrix $M \in \mathbf{R}^{(\max_{i\in\mathbf{v}} m_i)\times r}$ such that*

$$\begin{bmatrix} A-\lambda_i I & B \\ MC & 0 \end{bmatrix} = \textit{full rank}, \quad \forall i \in \mathbf{n}. \tag{8.6}$$

Consider a decentralized controller K_d with ν local controllers $K_{d_1}, K_{d_2}, ..., K_{d_\nu}$, where the local controller K_{d_i} ($\forall i \in \mathbf{v}$) receives $y_i(t)$ as its input to generate $u_i(t)$ as its output. Let K_d be an interconnection of two linear decentralized sub-controllers K_d^1 and K_d^2 with the sets of local controllers $\{K_{d_1}^1, K_{d_2}^1, ..., K_{d_\nu}^1\}$ and $\{K_{d_1}^2, K_{d_2}^2, ..., K_{d_\nu}^2\}$, respectively, such that $K_{d_i}^1$ generates the input signal $u_i(t)$ and another signal $y_{d_i}(t)$ in terms of $y_i(t)$ and $u_{d_i}(t)$, where $u_{d_i}(t)$ and $y_{d_i}(t)$ are the respective output and input of $K_{d_i}^2$, for every $i \in \mathbf{v}$. This implies that K_d can be regarded as a two-level decentralized controller with the high-level sub-controller K_d^1 and the low-level sub-controller K_d^2. The topology of the controller K_d is illustrated in Figure 8.1 for the particular case $\nu = 2$, which shows that the high-level controller K_d^1 interacts directly with the system $\mathscr{S}$ while the low-level controller K_d^2 can only communicate with the high-level controller K_d^1. Let the local controller $K_{d_i}^1$, $\forall i \in \mathbf{v}$, be as follows

$$\begin{bmatrix} \dot{x}_{d_i}(t) \\ \dot{z}_{d_i}(t) \end{bmatrix} = \begin{bmatrix} A+BD_oC & BC_o \\ B_oC & A_o \end{bmatrix} \begin{bmatrix} x_{d_i}(t) \\ z_{d_i}(t) \end{bmatrix} + \begin{bmatrix} B \\ 0 \end{bmatrix} u_{d_i}(t), \tag{8.7a}$$

$$y_{d_i}(t) = y_i(t) - \begin{bmatrix} C_i & 0 \end{bmatrix} \begin{bmatrix} x_{d_i}(t) \\ z_{d_i}(t) \end{bmatrix}, \tag{8.7b}$$

$$u_i(t) = \begin{bmatrix} 0_{m_i\times m_1} & \cdots & 0_{m_i\times m_{i-1}} & I_{m_i} & 0_{m_i\times m_{i+1}} & \cdots & 0_{m_i\times m_\nu} \end{bmatrix} \begin{bmatrix} D_oC & C_o \end{bmatrix} \begin{bmatrix} x_{d_i}(t) \\ z_{d_i}(t) \end{bmatrix}, \tag{8.7c}$$

where $x_{d_i}(t) \in \mathbf{R}^n$ and $z_{d_i}(t) \in \mathbf{R}^{n_o}$ together form the state vector of $K_{d_i}^1$ (note that $0_{\mu_1\times\mu_2}$ denotes a zero matrix of dimension $\mu_1 \times \mu_2$ in this chapter, for every natural numbers μ_1 and μ_2). To comprehend the basic idea behind the above control law, the configuration of the system $\mathscr{S}$ under the introduced decentralized controller K_d is sketched in Figure 8.2 for the particular case $\nu = 2$. Before designing the low-level decentralized sub-controller K_d^2, it is essential to understand why K_d^1 is defined as such.

Theorem 1 *Consider the decentralized controller K_d with the high-level sub-controller K_d^1 given in (8.7) and an arbitrary low-level linear sub-controller K_d^2. If the initial state of the high-level decentralized sub-controller K_d^1 is taken as*

$$\mathscr{Z}_0 := \Big[\underbrace{\begin{bmatrix} x_0^T & 0_{1\times n_0} \end{bmatrix} \begin{bmatrix} x_0^T & 0_{1\times n_0} \end{bmatrix} \cdots \begin{bmatrix} x_0^T & 0_{1\times n_0} \end{bmatrix}}_{\nu \text{ times}} \Big]^T \tag{8.8}$$

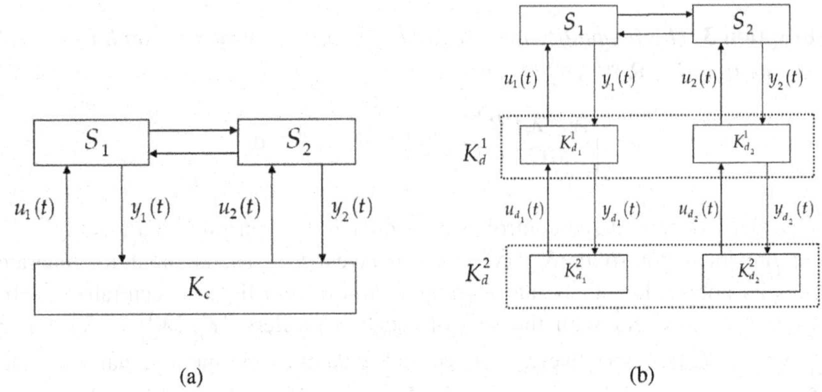

Fig. 8.1 (a): The block diagram of the system $\mathscr{S}$ under the centralized controller K_c (assuming $\nu = 2$); (b): The block diagram of the system $\mathscr{S}$ under the two-level decentralized controller K_d (assuming $\nu = 2$).

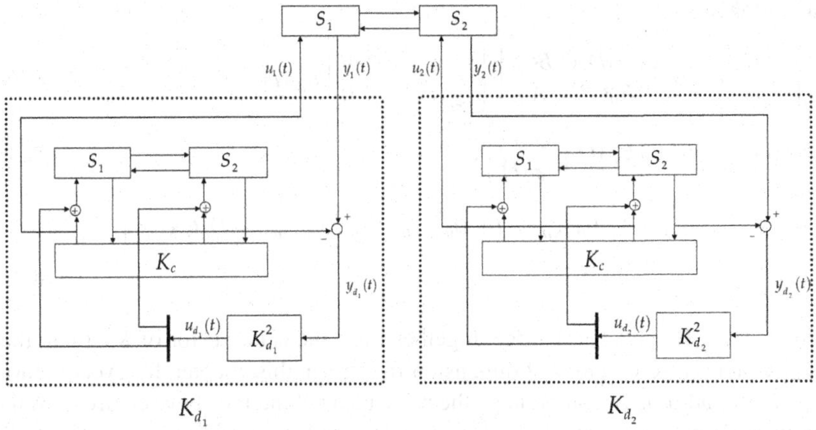

Fig. 8.2 The detailed block-diagram of the system under the decentralized controller K_d (assuming $\nu = 2$).

and the state of K_d^2 is initialized as zero, then the state and the input of the system $\mathscr{S}$ under the decentralized controller K_d will be identical to $x_c(t)$ and $u_c(t)$, respectively.

Proof: This theorem can be proved in line with the state-space approach proposed in the proof of Theorem 1 in [19]. However, an alternative method will be pursued here based on a well-known technique for circuit analysis. First, consider the system $\mathscr{S}$ under the high-level sub-controller K_d^1 and assume that the low-level sub-controller K_d^2 does not exist (i.e. $u_{d_1}(t) = \cdots = u_{d_\nu}(t) = 0$). Since $u_{d_1}(t), ..., u_{d_\nu}(t)$ are equal to zero for all $t \geq 0$, it follows from (8.7a), (8.7c) and the assumption

$$\left[x_{d_i}(0)^T \; z_{d_i}(0)^T\right] = \left[x_0^T \; 0_{1\times n_o}\right], \quad \forall i \in \nu \tag{8.9}$$

that

$$x_{d_i}(t) = x_c(t), \quad z_{d_i}(t) = z_c(t), \quad u_i(t) = u_{c_i}(t), \quad \forall t \geq 0, \, i \in \nu. \tag{8.10}$$

This means that the input $u(t)$ of the system $\mathscr{S}$ is equal to $u_c(t)$, which makes the state of the system be equal to $x_c(t)$, i.e.,

$$x(t) = x_c(t), \quad u(t) = u_c(t), \quad \forall t \geq 0. \tag{8.11}$$

The above relations yield that $y(t) = y_c(t)$, which can be combined with (8.7b) to obtain $y_{d_i}(t) = 0$ for every $i \in \nu$. Now, notice that the signals $u_{d_1}(t), ..., u_{d_\nu}(t)$ and $y_{d_1}(t), ..., y_{d_\nu}(t)$ are all equal to zero. As a result, if a <u>linear</u> controller K_d^2 with zero initial state is augmented to the system $\mathscr{S}$ under K_d^1 from the inputs $u_{d_1}(t), ..., u_{d_\nu}(t)$ to the outputs $y_{d_1}(t), ..., y_{d_\nu}(t)$, it will not change the zero values of these signals. In other words, the linear controller K_d^2 has no effect on the signals of the system $\mathscr{S}$ and the controller K_d^1. The proof now follows immediately from (8.11). ■

Given arbitrary vectors α, β, γ of appropriate dimensions, let $x_d(t; \alpha, \beta, \gamma)$ denote the state of the system $\mathscr{S}$ under the decentralized controller K_d provided the initial states of $\mathscr{S}$, K_d^1 and K_d^2 are taken as α, β and γ, respectively. Likewise, define the input signal $u_d(t; \alpha, \beta, \gamma)$. Theorem 1 states that

$$x_d(t; x_0, Z_0, 0) = x_c(t), \quad u_d(t; x_0, Z_0, 0) = u_c(t), \quad \forall t \geq 0, \tag{8.12}$$

for every arbitrary low-level linear sub-controller K_d^2. This implies that if the initial state of K_d^1 is taken as Z_0, then the left side of the inequality (8.4) is equal to zero for the above-defined controller K_d, which would make the centralized controller K_c decentrally implementable. Nevertheless, since the initial state x_0 is unknown by assumption and Z_0 is based on x_0, this initialization of K_d^1 is infeasible. Instead, the high-level controller K_d^1 can be initialized as zero, which leads to the state $x_d(t; x_0, 0, 0)$ and the input $u_d(t; x_0, 0, 0)$ for the system $\mathscr{S}$. Now, it follows from the linearity of the controller K_d and the relations given in (8.12) that

$$\begin{aligned}
&\int_{\varepsilon}^{\infty} \left(\|x_c(t) - x_d(t;x_0,0,0)\|^2 + \|u_c(t) - u_d(t;x_0,0,0)\|^2\right) dt = \\
&= \int_{\varepsilon}^{\infty} \left(\|x_d(t;x_0,Z_0,0) - x_d(t;x_0,0,0)\|^2 + \|u_d(t;x_0,Z_0,0) - u_d(t;x_0,0,0)\|^2\right) dt \\
&= \int_{\varepsilon}^{\infty} \left(\|x_d(t;0,Z_0,0)\|^2 + \|u_d(t;0,Z_0,0)\|^2\right) dt.
\end{aligned} \tag{8.13}$$

Hence, to prove that the centralized controller K_c is decentrally implementable for the system $\mathscr{S}$ with an unknown initial state x_0, it is enough to solve the next problem.

Problem 1: For every given positive numbers ε and Δ, design a low-level decentralized sub-controller K_d^2 such that when the initial states of $\mathscr{S}$, K_d^1 and K_d^2 are taken as 0, Z_0 and 0, respectively, then the state and input of the system $\mathscr{S}$ under the decentralized controller K_d satisfy the relation

$$\int_{\varepsilon}^{\infty} \left(\|x_d(t;0,Z_0,0)\|^2 + \|u_d(t;0,Z_0,0)\|^2\right) dt < \Delta. \tag{8.14}$$

Augment the system $\mathscr{S}$ with the high-level decentralized sub-controller K_d^1 to obtain an interconnected system $\tilde{\mathscr{S}}$ with ν subsystems, where the input and output of its ith subsystem are $u_{d_i}(t)$ and $y_{d_i}(t)$, respectively, for every $i \in \nu$ (this augmentation may be observed in Figure 8.1). Problem 1 stated above amounts to designing a linear decentralized controller K_d^2 for the system $\tilde{\mathscr{S}}$ with the initial state $\left[0_{1\times n}\ Z_0^T\right]^T$ such that the state of the system is regulated arbitrarily fast. Although a state regulation problem is normally easy-to-solve for a controllable and observable system, it will be shown in the sequel that the system $\tilde{\mathscr{S}}$ is not observable.

Denote the eigenvalues of the centralized closed-loop system (i.e. the system $\mathscr{S}$ under the controller K_c) with $\bar{\lambda}_1, \bar{\lambda}_2, \dots, \bar{\lambda}_{n+n_o}$. Besides, let Φ be an $(n+n_o)$-dimensional subspace defined as

$$\Phi := \left\{ \Big[\eta\ \underbrace{[\eta\ \gamma]\ [\eta\ \gamma]\ \cdots\ [\eta\ \gamma]}_{\nu \text{ times}}\Big]^T \,\Big|\, \eta \in \mathbf{R}^{1\times n},\ \gamma \in \mathbf{R}^{1\times n_o} \right\} \tag{8.15}$$

Theorem 2 *There exists a nonzero multivariate polynomial* $F : \mathbf{R}^{n_o\times n_o} \times \mathbf{R}^{n_o\times r} \times \mathbf{R}^{m\times n_o} \times \mathbf{R}^{m\times r} \to \mathbf{R}$ *(whose coefficients are dependent only on the parameters of the system* $\mathscr{S}$ *rather than* K_c*) such that the following statements hold if* $F(A_o, B_o, C_o, D_o) \neq 0$*:*

i) The system $\tilde{\mathscr{S}}$ *is strongly connected.*
ii) The modes of the system $\tilde{\mathscr{S}}$ *are*

$$\left(\bigcup_{j=1}^{n+n_o} \{\underbrace{\bar{\lambda}_j, \dots, \bar{\lambda}_j}_{\nu \text{ times}}\} \right) \bigcup \{\lambda_1, \dots, \lambda_n\} \tag{8.16}$$

(the multiplicities of λ_i and $\bar{\lambda}_j$ in the above set are 1 and ν, respectively, for every $i \in \mathbf{n}$ and $j \in \{1,2,...,n+n_o\}$).

iii) The system $\tilde{\mathscr{S}}$ is controllable.

iv) The system $\tilde{\mathscr{S}}$ is unobservable with the unobservable subspace Φ corresponding to the $n+n_0$ unobservable modes $\bar{\lambda}_1,...,\bar{\lambda}_{n+n_o}$ (each of these repeated modes is only one time unobservable).

v) The mode λ_j is not a DFM of the system $\tilde{\mathscr{S}}$, for every $j \in \mathbf{n}$.

vi) The repeated mode $\bar{\lambda}_j$ with multiplicity ν is a single DFM of the system $\tilde{\mathscr{S}}$, for every $j \in \{1,2,...,n+n_o\}$.

Proof: The proof of this theorem is provided in Appendix 1. ■

Several properties of the system $\tilde{\mathscr{S}}$ have been derived in Theorem 2 under the condition $F(A_o,B_o,C_o,D_o) \neq 0$. A question arises: how can the validity of this condition be checked? A method is proposed in Appendix 2 to find the multivariate polynomial $F(A_o,B_o,C_o,D_o)$. However, this polynomial has a complicated structure with several monomials, in general. Hence, it may not be useful to find this polynomial explicitly. In contrast, one can argue that given a fixed number n_o, almost all (generic) finite-dimensional LTI controllers K_c satisfy the property $F(A_o,B_o,C_o,D_o) \neq 0$; more precisely, the set of controllers K_c for which this condition is violated forms a set of measure zero. The reason is that the set of real zeros of a (nonzero) polynomial is a hypersurface (real algebraic variety) with a positive codimension. Therefore, instead of checking the condition $F(A_o,B_o,C_o,D_o) \neq 0$ directly for a given controller K_c, it is easier to verify whether properties (i)-(vi) stated in Theorem 2 hold. Based on the aforementioned discussion, if any of these properties is violated, then one can randomly perturb the parameters A_0,B_0,C_0,D_o arbitrarily small to obtain a perturbed controller for which the properties given in Theorem 2 hold with probability 1.

Assume henceforth that the condition $F(A_o,B_o,C_o,D_o) \neq 0$ is satisfied for the centralized controller K_c. To design the low-level decentralized sub-controller K_d^2, let its first $\nu-1$ local controllers be simply static feedbacks, while its last local controller may possibly be a dynamic controller. For this purpose, given arbitrary gains $G_i \in \mathbf{R}^{m\times r_i}$, $\forall i \in \{1,2,...,\nu-1\}$, let $\tilde{\mathscr{S}}(G_1,...,G_{\nu-1})$ denote the system $\tilde{\mathscr{S}}$ under the following local controllers $K_{d_1}^2,...,K_{d_{\nu-1}}^2$:

$$u_{d_i}(t) = G_i y_{d_i}(t), \quad \forall i \in \{1,2,...,\nu-1\}, \tag{8.17}$$

where the input and output of $\tilde{\mathscr{S}}(G_1,...,G_{\nu-1})$ are $u_{d_\nu}(t)$ and $y_{d_\nu}(t)$, respectively.

Theorem 3 *There exists a multivariate polynomial $\bar{F}: \mathbf{R}^{m\times r_1} \times \cdots \times \mathbf{R}^{m\times r_{\nu-1}} \to \mathbf{R}$ such that for every given gains $G_i \in \mathbf{R}^{m\times r_i}$, $\forall i \in \{1,2,...,\nu-1\}$, if $\bar{F}(G_1,G_2,...,G_{\nu-1}) \neq 0$, then the system $\tilde{\mathscr{S}}(G_1,...,G_{\nu-1})$ satisfies the following properties:*

i) $\tilde{\mathscr{S}}(G_1,...,G_{\nu-1})$ is controllable.

ii) $\tilde{\mathscr{S}}(G_1,...,G_{\nu-1})$ is unobservable with the unobservable subspace Φ.

Proof: Since the system $\tilde{\mathscr{S}}$ is unobservable (due to Theorem 2), consider a minimal realization of this system and denote it with $\bar{\mathscr{S}}$. Define the system $\bar{\mathscr{S}}(G_1,...,$

$G_{\nu-1})$ in the same way that $\tilde{\mathscr{S}}(G_1,...,G_{\nu-1})$ was defined. It follows from Theorem 2 that the interconnected system $\bar{\mathscr{S}}$ is strongly connected, controllable, observable, and has no DFMs. Hence, it can be concluded from [31] and [32] that there exists a polynomial $\bar{F}: \mathbf{R}^{m\times r_1} \times \cdots \times \mathbf{R}^{m\times r_{\nu-1}} \to \mathbf{R}$ such that the system $\bar{\mathscr{S}}(G_1,...,G_{\nu-1})$ is both controllable and observable for every set of gains $\{G_1,G_2,...,G_{\nu-1}\}$ satisfying the relation $\bar{F}(G_1,G_2,...,G_{\nu-1}) \neq 0$. The proof of this theorem is a consequence of this property of the system $\bar{\mathscr{S}}(G_1,...,G_{\nu-1})$ and the following facts:

- $\bar{\mathscr{S}}$ is a reduced-order observable realization of the system $\tilde{\mathscr{S}}$.
- The unobservable subspace of the system $\tilde{\mathscr{S}}$ is shown in Theorem 2 to be Φ. ■

The discussion given right after Theorem 2 concerning the polynomial F is applicable to the polynomial $\bar{F}$ as well. In other words, even though the polynomial $\bar{F}$ can be found explicitly, one can argue that the condition $\bar{F}(G_1,G_2,...,G_{\nu-1}) \neq 0$ holds for a generic choice of gains $G_1,G_2,...,G_{\nu-1}$. From now on, let $\{G_1,G_2,...,G_{\nu-1}\}$ denote a specific set of gains for which properties (i) and (ii) mentioned in Theorem 3 hold for the system $\tilde{\mathscr{S}}(G_1,...,G_{\nu-1})$.

Recall from Problem 1 and its subsequent discussion that the decentralized implementability of the centralized controller K_c is guaranteed by designing a local controller $K_{d_\nu}^2$ for the system $\tilde{\mathscr{S}}(G_1,...,G_{\nu-1})$ with the initial state $\left[0_{1\times n}\ Z_0^T\right]^T$ so that the state of the closed-loop system is regulated arbitrarily fast. To address the latter problem, note that although the system $\tilde{\mathscr{S}}(G_1,...,G_{\nu-1})$ is not observable (due to Theorem 3), the unobservable modes of the system are all stable. This implies that one can design an observer-based controller $K_{d_\nu}^2$ for the system $\tilde{\mathscr{S}}(G_1,...,G_{\nu-1})$ to make its state attenuate to zero. However, the question of interest is to design an *arbitrarily fast* regulating controller for this unobservable system. This question is addressed in the sequel through the introduction of a new notion.

8.3.1 Structural initial value observer

Consider an LTI continuous-time system $\mathbf{S}$ with the state-space representation

$$\begin{aligned}\dot{\mathbf{x}}(t) &= \mathbf{A}\mathbf{x}(t) + \mathbf{B}\mathbf{u}(t),\\ \mathbf{y}(t) &= \mathbf{C}\mathbf{x}(t),\end{aligned} \tag{8.18}$$

where $\mathbf{x}(t) \in \mathbf{R}^{\bar{n}}$, $\mathbf{u}(t) \in \mathbf{R}^{\bar{m}}$ and $\mathbf{y}(t) \in \mathbf{R}^{\bar{r}}$. Assume that Γ is a given subspace of $\mathbf{R}^{\bar{n}}$.

Definition 2 *An initial state $\mathbf{x}_0 \in \Gamma$ is said to be structural initial value observable with respect to Γ (SIV observable w.r.t. Γ) if there does not exist another initial state $\mathbf{x}_0'$ in the space Γ such that the system $\mathbf{S}$ results in the same output by starting from each of the initial states $\mathbf{x}_0$ and $\mathbf{x}_0'$. Furthermore, the system $\mathbf{S}$ is called SIV observable w.r.t. Γ if every initial state $\mathbf{x}_0$ in Γ is SIV observable w.r.t. Γ.*

It is evident that the system $\mathbf{S}$ is SIV observable w.r.t. Γ if and only if the intersection of Γ and the unobservable subspace of $\mathbf{S}$ is only the origin. Assume that $\mathbf{S}$ is controllable, unobservable with an unobservable subspace of dimension $\mu \in \mathbf{N}$, and SIV observable w.r.t. Γ. A question of interest is how to design an arbitrarily fast observer for this system if it is known *a priori* that the initial state of the system belongs to Γ. To address this question, realize the system in the Kalman observable form as

$$\begin{bmatrix} \dot{\mathbf{z}}_1(t) \\ \dot{\mathbf{z}}_2(t) \end{bmatrix} = \begin{bmatrix} \mathbf{A}_{11} & 0 \\ \mathbf{A}_{21} & \mathbf{A}_{22} \end{bmatrix} \begin{bmatrix} \mathbf{z}_1(t) \\ \mathbf{z}_2(t) \end{bmatrix} + \begin{bmatrix} \mathbf{B}_1 \\ \mathbf{B}_2 \end{bmatrix} \mathbf{u}(t), \qquad \mathbf{y}(t) = \mathbf{C}_1 \mathbf{z}_1(t), \tag{8.19}$$

where $\mathbf{A}_{22} \in \mathbf{R}^{\mu \times \mu}$, the pair $(\mathbf{A}_{11}, \mathbf{C}_1)$ is observable, and the relations $\mathbf{z}_1(t) = \mathbf{T}_1 \mathbf{x}(t)$ and $\mathbf{z}_2(t) = \mathbf{T}_2 \mathbf{x}(t)$ hold for some appropriate similarity transformation matrix $\mathbf{T} := [\mathbf{T}_1^T \ \mathbf{T}_2^T]^T$. For simplicity, assume that $\mathbf{A}_{22}$ is a Hurwitz matrix, i.e., all unobservable modes of the systems are stable (this condition holds for the decentralized problem under investigation in this chapter).

Consider an initial state $\mathbf{x}[0] \in \Gamma$. Notice that $\mathbf{z}_2(t)$ cannot be observed in the output $\mathbf{y}(t)$ and, on the other hand, since the system is SIV observable w.r.t. Γ, the signal $\mathbf{z}_2(t)$ should be recoverable from the observable state $\mathbf{z}_1(t)$. Indeed, it can be verified that there exists a linear map $\zeta(\cdot)$ (matrix) such that $\mathbf{z}_2(0) = \zeta \mathbf{z}_1(0)$ for every $\mathbf{x}_0 \in \Gamma$, where $\mathbf{z}_1(0) = \mathbf{T}_1 \mathbf{x}_0$ and $\mathbf{z}_2(0) = \mathbf{T}_2 \mathbf{x}_0$.

It can be verified that an arbitrarily fast observer in the standard form, i.e. a Luenberger observer, may not exist if $\bar{r} < \mu$, due to the presence of an unobservable subspace of dimension μ. Nonetheless, it may be speculated that one can simply design an observer to recover $\mathbf{z}_1(t)$ from the observed output $\mathbf{y}(t)$ and then design a compensator to retrieve $\mathbf{z}_2(t)$ from $\mathbf{z}_1(t)$. This idea normally fails for designing an arbitrarily fast observer. The reason is that a fast observer for $\mathbf{z}_1(t)$ often leads to a large overshoot in the estimation of $\mathbf{z}_1(t)$ and since $\mathbf{z}_2(t)$ must be retrieved from $\mathbf{z}_1(t)$ using an open-loop compensator (based on the linear map ζ), there is no way to diminish the effect of overshoot quickly in the estimation of $\mathbf{z}_2(t)$. To this end, a more complex observer will be introduced in the sequel. Select a positive real τ and define

$$\Pi := e^{\mathbf{A}_{22}\tau} \zeta e^{-\mathbf{A}_{11}\tau} + \int_0^{\tau} e^{\mathbf{A}_{22}(\tau - t)} \mathbf{A}_{21} e^{-\mathbf{A}_{11}(\tau - t)} dt. \tag{8.20}$$

Consider the transfer function $(sI - \mathbf{A}_{22})^{-1}(\mathbf{A}_{21} + \Pi(sI - \mathbf{A}_{11}))$. Since this is a proper function, it can be realized in the space-state form. Let a realization of this transfer function be given by the state-space matrices $(\mathbf{M}_1, \mathbf{M}_2, \mathbf{M}_3, \mathbf{M}_4)$. Due to the stability of the matrix $\mathbf{A}_{22}$, the matrix M_1 must be Hurwitz.

Theorem 4 *Assume that the system* $\mathbf{S}$ *starts from an unknown initial state* $\mathbf{x}_0 \in \Gamma$. *Given a matrix* L_1 *such that* $\mathbf{A}_{11} + L_1 \mathbf{C}_1$ *is Hurwitz, consider the compensator*

$$\dot{\hat{\mathbf{z}}}_1(t) = \mathbf{A}_{11}\hat{\mathbf{z}}_1(t) + \mathbf{B}_1\mathbf{u}(t) + L_1(\mathbf{C}_1\hat{\mathbf{z}}_1(t) - \mathbf{y}(t)), \quad \hat{\mathbf{z}}_1(0) = 0, \tag{8.21a}$$
$$\dot{\mathbf{z}}_{11}(t) = \mathbf{A}_{11}\mathbf{z}_{11}(t) + \mathbf{B}_1\mathbf{u}(t), \quad \mathbf{z}_{11}(0) = 0, \tag{8.21b}$$
$$\dot{\mathbf{z}}_{21}(t) = \mathbf{A}_{21}\mathbf{z}_{11}(t) + \mathbf{A}_{22}\mathbf{z}_{21}(t) + \mathbf{B}_2\mathbf{u}(t), \quad \mathbf{z}_{21}(0) = 0, \tag{8.21c}$$
$$\dot{\mathbf{z}}_p(t) = \mathbf{M}_1\mathbf{z}_p(t) + \mathbf{M}_2(\hat{\mathbf{z}}_1(t) - \mathbf{z}_{11}(t))u_s(t-\tau), \quad \mathbf{z}_p(0) = 0, \tag{8.21d}$$
$$\mathbf{y}_p(t) = \mathbf{M}_3\mathbf{z}_p(t) + \mathbf{M}_4(\hat{\mathbf{z}}_1(t) - \mathbf{z}_{11}(t))u_s(t-\tau), \tag{8.21e}$$
$$\hat{\mathbf{z}}_2(t) = \mathbf{z}_{21}(t) + \mathbf{y}_p(t), \tag{8.21f}$$

where $u_s(\cdot)$ is the step function. This is an SIV observer for the system $\mathbf{S}$ *(where $\hat{\mathbf{z}}_i$ estimates $\mathbf{z}_i$ for $i = 1,2$), which satisfies the following properties:*

i) It is internally stable.
ii) The state estimation error converges to zero.
iii)The state estimation error is independent of the input of the system.

In addition, the observation process can be made arbitrarily fast by letting τ go to zero and pushing the eigenvalues of $\mathbf{A}_{11} + L_1\mathbf{C}_1$ towards $-\infty$.

Proof: It is evident that $\hat{\mathbf{z}}_1(t) \to \mathbf{z}_1(t)$, as t goes to infinity. On the other hand, one can write $\mathbf{z}_1(t) = \mathbf{z}_{11}(t) + \mathbf{z}_{12}(t)$, where $\mathbf{z}_{11}(t)$ is given by the differential equation (8.21b) and

$$\dot{\mathbf{z}}_{12}(t) = \mathbf{A}_{11}\mathbf{z}_{12}(t), \quad \mathbf{z}_{12}(0) = \mathbf{z}_1(0). \tag{8.22}$$

Furthermore, $\mathbf{z}_2(t)$ can be decomposed as $\mathbf{z}_{21}(t) + \mathbf{z}_{22}(t)$, where $\mathbf{z}_{21}(t)$ is given by (8.21c) and

$$\dot{\mathbf{z}}_{22}(t) = \mathbf{A}_{21}\mathbf{z}_{12}(t) + \mathbf{A}_{22}\mathbf{z}_{22}(t), \quad \mathbf{z}_{22}(0) = \mathbf{z}_2(0). \tag{8.23}$$

Let $\mathscr{L}\{\cdot\}$ represent the Laplace transformation. It holds that

$$\begin{aligned} \mathscr{L}\{\mathbf{z}_{22}(t)u_s(t-\tau)\} &= (sI - \mathbf{A}_{22})^{-1}\mathbf{A}_{21}\mathscr{L}\{\mathbf{z}_{12}(t)u_s(t-\tau)\} \\ &\quad + e^{-\tau s}(sI - \mathbf{A}_{22})^{-1}\mathbf{z}_{22}(\tau). \end{aligned} \tag{8.24}$$

Besides

$$\begin{aligned} \mathbf{z}_{22}(\tau) &= e^{\mathbf{A}_{22}\tau}\mathbf{z}_{22}(0) + \int_0^\tau e^{\mathbf{A}_{22}(\tau-t)}\mathbf{A}_{21}\mathbf{z}_{12}(t)dt \\ &= \left(e^{\mathbf{A}_{22}\tau}\zeta + \int_0^\tau e^{\mathbf{A}_{22}(\tau-t)}\mathbf{A}_{21}e^{\mathbf{A}_{11}t}dt\right)\mathbf{z}_1(0) = \Pi\mathbf{z}_{12}(\tau) \end{aligned} \tag{8.25}$$

(note that the last line of the above equation is a consequence of (8.22)). Moreover

$$\mathbf{z}_{12}(\tau) = e^{\tau s}(sI - \mathbf{A}_{11})\mathscr{L}\{\mathbf{z}_{12}(t)u_s(t-\tau)\}. \tag{8.26}$$

Thus, it results from the equations (8.24), (8.25) and (8.26) that

$$\mathscr{L}\{\mathbf{z}_{22}(t)u_s(t-\tau)\} = (sI - \mathbf{A}_{22})^{-1}(\mathbf{A}_{21} + \Pi(sI - \mathbf{A}_{11}))\mathscr{L}\{\mathbf{z}_{12}(t)u_s(t-\tau)\} \tag{8.27}$$

or equivalently

$$\begin{aligned}\dot{\mathbf{z}}_o(t) &= \mathbf{M}_1\mathbf{z}_o(t) + \mathbf{M}_2\mathbf{z}_{12}(t)u_s(t-\tau),\\ \mathbf{z}_{22}(t)u_s(t-\tau) &= \mathbf{M}_3\mathbf{z}_o(t) + \mathbf{M}_4\mathbf{z}_{12}(t)u_s(t-\tau),\end{aligned} \tag{8.28}$$

for some state $\mathbf{z}_o$. In order to prove that the compensator (8.21) is an SIV observer, it is enough to notice that $(\hat{\mathbf{z}}_1(t) - \mathbf{z}_{11}(t))\,u_s(t-\tau) \rightarrow \mathbf{z}_{12}(t)u_s(t-\tau)$, as $t \rightarrow \infty$. Now, it is straightforward to show that Properties (i), (ii) and (iii) hold (notice that the eigenvalues of the matrix $\mathbf{M}_1$ are identical to those of the matrix $\mathbf{A}_{22}$, which are all stable by assumption). This observer can be made arbitrarily fast in light of the following facts:

- $\mathbf{z}_1(t)$ can be recovered arbitrarily fast by means of a proper high gain matrix L_1.
- Although making the recovery process of $\mathbf{z}_1(t)$ fast would result in a large overshoot, its deteriorating effect can be nullified by the term $u_s(t-\tau)$, which discards some undesirable part of the signal $\hat{\mathbf{z}}_{11}(t)$ (see the equations (8.21d) and (8.21e)). Thus, it is essential that τ be positive, and clearly $\tau = 0$ may not lead to an arbitrarily fast observer for $\mathbf{z}_2(t)$. ■

Corollary 1 *Assume that the system* $\mathbf{S}$ *starts from an unknown initial state* $\mathbf{x}_0 \in \Gamma$. *Given matrix gains* L_1 *and* Q *(with appropriate dimensions) such that* $\mathbf{A}_{11} + L_1\mathbf{C}_1$ *and* $\mathbf{A} + \mathbf{B}Q$ *are both Hurwitz, consider the system* $\mathbf{S}$ *under an observer-based controller composed of the SIV observer (8.21) and the static controller* $\mathbf{u}(t) = Q\mathbf{T}^{-1}\left[\hat{\mathbf{z}}_1(t)^T \ \hat{\mathbf{z}}_2(t)^T\right]^T$. *The closed-loop system is stable and, more precisely, the state of the closed-loop system can be pushed towards zero arbitrarily fast by making the eigenvalues of* $\mathbf{A}_{11} + L_1\mathbf{C}_1$ *and* $\mathbf{A} + \mathbf{B}Q$ *go to* $-\infty$ *and letting the parameter* τ *go to zero.*

Proof: Since the SIV observer proposed in Theorem 4 satisfies Properties (i)-(iii) stated there, the proof follows immediately from the fact that the well-known separation principle holds for the observer-based controller given in the corollary (note that the signal $\mathbf{T}^{-1}\left[\hat{\mathbf{z}}_1(t)^T \ \hat{\mathbf{z}}_2(t)^T\right]^T$ estimates the state $\mathbf{x}(t)$). ■

8.3.2 Decentralization of centralized controllers

For a given controller K_c, assume that $F(A_o, B_o, C_o, D_o) \neq 0$. Consider a decentralized controller K_d consisting of a high-level decentralized controller K_d^1 given in (8.7) and a low-level decentralized controller K_d^2 with its first $\nu - 1$ local controllers being static as given in (8.17) such that $\bar{F}(G_1, G_2, ..., G_{\nu-1}) \neq 0$. Augment the system $\mathscr{S}$ with the high-level controller K_d^1 and these $\nu - 1$ low-level local controllers to obtain a system $\bar{\mathscr{S}}(G_1, G_2, ..., G_{\nu-1})$. Let $\bar{\Phi}$ be an n-dimensional subspace defined as

$$\bar{\Phi} := \left\{ \left[0_{1\times n} \ \underbrace{[\eta \ 0_{1\times n_o}] \ [\eta \ 0_{1\times n_o}] \ \cdots \ [\eta \ 0_{1\times n_o}]}_{\nu \text{ times}} \right]^T \,\middle|\, \eta \in \mathbf{R}^{1\times n} \right\} \tag{8.29}$$

As discussed after Theorem 3, the "decentralized implementation" of K_c amounts to designing an arbitrarily fast state-regulating controller $K^2_{d_v}$ for the controllable but unobservable system $\tilde{\mathscr{S}}(G_1, G_2, ..., G_{v-1})$ with the *artificial* initial state $\left[0_{1\times n}\ Z_0^T\right]^T \in \bar{\Phi}$. Since the intersection of the initial-state subspace $\bar{\Phi}$ with the unobservable subspace Φ (see Theorem 3) is only the origin, the system $\tilde{\mathscr{S}}(G_1, G_2, ..., G_{v-1})$ is SIV observable w.r.t. $\bar{\Phi}$. Hence, by taking $\mathbf{S}$ as $\tilde{\mathscr{S}}(G_1, G_2, ..., G_{v-1})$ and $\mathbf{x}_0$ as $\left[0_{1\times n}\ Z_0^T\right]^T$, Corollary 1 can be exploited to design an arbitrarily fast state-regulating observed-based controller $K^2_{d_v}$ for $\tilde{\mathscr{S}}(G_1, G_2, ..., G_{v-1})$, which is composed of an SIV observer w.r.t. $\bar{\Phi}$ and a static controller. Note that the parameter ξ introduced in Definition 1 is equal to (τ, L_1, Q), where τ, L_1, Q are the parameters of the observed-based controller $K^2_{d_v}$.

Remark 1 *It can be deduced from the equation (8.10) in the proof of Theorem 1 and the observer-based nature of the low-level sub-controller K^2_d that $x_{d_i}(t) \to x_c(t)$ as t tends to infinity. This implies that each local controller for the system $\mathscr{S}$ is equipped with an internal observer to asymptotically estimate the global state of the system.*

Remark 2 *A question arises as whether ε in Definition 1 can be set to zero or it must be strictly positive. Based on the proposed formulation, this question amounts to checking the perfect regulation and bounded peaking properties of the system $\tilde{S}(G_1, G_2, ..., G_{v-1})$. If this system possesses these properties, then ε can be taken as zero. For a detailed discussion on perfect regulation, one can refer to [34].*

Remark 3 *Although this work develops a method for the decentralized implementation of a centralized controller, it can be easily extended to the decentralized overlapping implementation of a centralized controller as well. This can be carried out using the bijective transformation given in [33] between decentralized overlapping and decentralized control systems.*

8.3.3 Optimal decentralized controller

The objective of this part is to study how the results of the present work can be exploited to tackle the important problem of the optimal decentralized controller design. To this end, consider a simple interconnected system $\mathscr{S}$ with two subsystems characterized by the parameters

$$A = \begin{bmatrix} 8 & 1 \\ -8 & -2 \end{bmatrix}, \quad B_1 = C_1^T = \begin{bmatrix} 1 \\ 0 \end{bmatrix}, \quad B_2 = C_2^T = \begin{bmatrix} 0 \\ 1 \end{bmatrix} \tag{8.30}$$

and the initial state $x_0 = \left[x_1(0)\ x_2(0)\right]^T$. Let the centralized controller K_c be given as

$$u(t) = \begin{bmatrix} -15 & 9 \\ 9 & -6 \end{bmatrix} y(t). \tag{8.31}$$

This controller minimizes the performance index

$$J = \int_0^\infty \left(x(t)^T Q x(t) + u(t)^T R u(t) \right) dt \tag{8.32}$$

for the system $\mathscr{S}$, where

$$Q = \begin{bmatrix} 0.0063 & -0.0793 \\ -0.0793 & 0.9937 \end{bmatrix}, \quad R = \begin{bmatrix} 0.0826 & 0.1264 \\ 0.1264 & 0.2090 \end{bmatrix}. \tag{8.33}$$

Consider the problem of designing an affine time-invariant decentralized controller K_d to minimize the cost function J under the assumption that each local station (controller) knows the initial state of its own subsystem, but does not know the initial state of the other subsystem (for instance, the local controller of S_1 can be designed in terms of $x_1(0)$, but $x_2(0)$ is unknown for this local controller). Define J_c and J_d as the optimal performance indices in the centralized and decentralized cases, respectively. Due to the uniqueness of the solution of the underlying Riccati equation in this example and the assumption that the local controller of the first subsystem does not know $x_2(0)$, it is easy to verify that there exists no affine time-invariant decentralized controller that is able to result in the performance index J_c. Thus, one would speculate that $J_d > J_c$. However, the goal is to show that $J_d = J_c$, although there exists no affine decentralized controller with finite parameters such that its corresponding cost is equal to J_d. In other words, it is desired to prove that there is a sequence of decentralized controllers whose performance indices converge to J_c.

To prove the above-mentioned statement, consider a decentralized controller K_d composed of two interconnected sub-controllers K_d^1 and K_d^2, where K_d^1 is given in (8.7) and K_d^2 is a static decentralized controller as

$$u_{d_1}(t) = \left[\xi \ \xi^2 \right]^T y_{d_1}(t), \quad u_{d_2}(t) = \left[-\xi^2 \ \xi \right]^T y_{d_2}(t), \tag{8.34}$$

where ξ is a scalar tuning parameter that will be specified later. In light of Theorem 1, if the sub-controllers $K_{d_1}^1$ and $K_{d_2}^1$ are initialized as

$$x_{d_1}(0) = x_{d_2}(0) = \left[x_1(0) \ x_2(0) \right]^T, \tag{8.35}$$

then the controllers K_d and K_c both generate the same input and state for the system $\mathscr{S}$. Nonetheless, since it is already assumed that the first local controller cannot use $x_2(0)$ in light of the interconnected structure of the system, this initialization is impossible. Assume that $K_{d_1}^1$ and $K_{d_2}^1$ are initialized as

$$x_{d_1}(0) = \left[x_1(0) \ 0 \right]^T, \quad x_{d_2}(0) = \left[0 \ x_2(0) \right]^T. \tag{8.36}$$

Let $x_d(t)$ and $u_d(t)$ denote the state and the input of the system $\mathscr{S}$ under the controller K_d initialized as above. Having defined $\tilde{\mathscr{S}}$ as the system $\mathscr{S}$ under K_d^1, denote the set of state-space matrices of this resultant system with $(\tilde{A}, \tilde{B}, \tilde{C})$ and its state with $\tilde{x}(t)$. Define also

$$\tilde{A}(\xi) = \tilde{A} + \tilde{B}\begin{bmatrix} \xi & \xi^2 & 0 & 0 \\ 0 & 0 & -\xi^2 & \xi \end{bmatrix}^T \tilde{C}, \quad \forall \xi \in \mathbf{R}. \tag{8.37}$$

Using the method proposed in [35] for checking the robust stability of a polynomially uncertain matrix, it can be shown that the matrix $\tilde{A}(\xi)$ is Hurwitz for every $\xi \geq 10$. On the other hand, the argument leading to the equation (8.13) can be adopted to conclude that $x_d(t) - x_c(t)$ and $u_d(t) - u_c(t)$ are equal to the state and input of the system $\mathscr{S}$ under the controller K_d, respectively, if the initial state of closed-loop system is taken as

$$\bar{x}_0 := \begin{bmatrix} 0 \; 0 \; 0 \; x_2(0) \; x_1(0) \; 0 \end{bmatrix}^T. \tag{8.38}$$

Therefore, the relation

$$\int_0^{\infty} \|x_d(t) - x_c(t)\|^2 dt \leq \int_0^{\infty} \|\tilde{x}(t)\|^2 dt = \text{trace}\left(\bar{x}_0^T P(\xi) \bar{x}_0\right) \tag{8.39}$$

holds for every $\xi \geq 10$, where

$$\tilde{A}(\xi)^T P(\xi) + P(\xi)\tilde{A}(\xi) + I = 0. \tag{8.40}$$

Due to the relation (8.39) and the particular structure of $\bar{x}_0$, in order to prove that $\int_0^\infty \|x_d(t) - x_c(t)\|^2 dt$ goes to zero as ξ tends to infinity, it suffices to show that the $(4,4)$, $(4,5)$ and $(5,5)$ entries of $P(\xi)$ all attenuate to zero as ξ goes to infinity. To prove this statement, one can solve the Lyapunov equation (8.40) using the well-known Kronecker-product method to deduce that every entry of $P(\xi)$ is a rational function in ξ (see the proof of Lemma 2 in [36]). Since the $(4,4)$, $(4,5)$ and $(5,5)$ entries of $P(\xi)$ all turn out to be strictly proper rational functions, they attenuate to zero as ξ goes to infinity. This yields

$$\int_0^{\infty} \|x_d(t) - x_c(t)\|^2 dt \to 0 \quad \text{as} \quad \xi \to \infty, \tag{8.41}$$

or more precisely

$$\int_0^{\infty} \|\tilde{x}(t)\|^2 dt \to 0 \quad \text{as} \quad \xi \to \infty. \tag{8.42}$$

The above relation can be combined with (8.7c) to obtain

$$\int_0^{\infty} \|u_d(t) - u_c(t)\|^2 dt \to 0 \quad \text{as} \quad \xi \to \infty. \tag{8.43}$$

It can be concluded from (8.41) and (8.43) that K_d parameterized in terms of ξ results in a performance index J arbitrarily close to J_c (by letting ξ go to infinity).

8.4 Numerical example

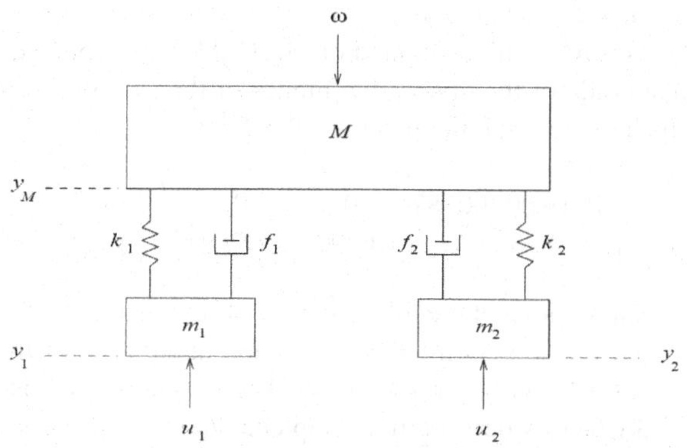

Fig. 8.3 The mass-spring system studied in the numerical example.

Consider the mass-spring system $\mathscr{S}$ given in Figure 8.3. Regard this system as a two-channel interconnected system with the input $u_i(t)$ and the output $y_i(t)$ for its ith control channel, where $i = 1,2$. By defining the state

$$x(t) := [y_M(t)\ \dot{y}_M(t)\ y_1(t)\ \dot{y}_1(t)\ y_2(t)\ \dot{y}_2(t)]^T, \tag{8.44}$$

the state-space matrices of this system (for the nominal values given in [37]) can be obtained as

$$A = \begin{bmatrix} 0 & 1 & 0 & 0 & 0 & 0 \\ -0.2 & -0.02 & 0.1 & 0.01 & 0.1 & 0.01 \\ 0 & 0 & 0 & 1 & 0 & 0 \\ 1 & 0.1 & -1 & -0.1 & 0 & 0 \\ 0 & 0 & 0 & 0 & 0 & 1 \\ 1 & 0.1 & 0 & 0 & -1 & -0.1 \end{bmatrix},$$
$$B_1 = \begin{bmatrix} 0\,0\,0\,1\,0\,0 \end{bmatrix}^T, \quad B_2 = \begin{bmatrix} 0\,0\,0\,0\,0\,1 \end{bmatrix}^T,$$
$$C_1 = \begin{bmatrix} 0\,0\,1\,0\,0\,0 \end{bmatrix}, \quad C_2 = \begin{bmatrix} 0\,0\,0\,0\,1\,0 \end{bmatrix}. \tag{8.45}$$

Assume that K_c is a centralized controller with the control law

$$\begin{bmatrix} u_1(t) \\ u_2(t) \end{bmatrix} = \begin{bmatrix} -3 & 2 \\ 3 & -5 \end{bmatrix} \begin{bmatrix} y_1(t) \\ y_2(t) \end{bmatrix}, \tag{8.46}$$

which is desired to be implemented in a decentralized fashion. Note that this controller results in a slow, oscillatory behavior for the system $\mathscr{S}$. Consider a decentralized controller K_d comprising two interconnected decentralized sub-controllers K_d^1 and K_d^2, where K_d^1 is given in (8.7). Recall that the present work suggests designing the low-level decentralized sub-controller K_d^2 in such a way that its first local controller is static while its second local controller is possibly dynamic (observer-based). However, let the possibility of designing a static low-level sub-controller K_d^2 be checked first. As stated in Theorem 2, the system $\mathscr{S}$ under K_d^1, denoted by $\tilde{\mathscr{S}}$, has 6 unobservable modes that are identical to the modes of the system $\mathscr{S}$ under the controller K_c. The "fminsearch" command of MATLAB was deployed to minimize the maximum magnitude of the observable modes of the system $\tilde{\mathscr{S}}$ under a *static* controller K_d^2, which led to a stabilizing controller K_d^2 as

$$\begin{bmatrix} u_{d_1}(t) \\ u_{d_2}(t) \end{bmatrix} = \begin{bmatrix} 10.0000 & 3.0187 & 0 & 0 \\ 0 & 0 & 10.0000 & 20.8750 \end{bmatrix}^T \begin{bmatrix} y_{d_1}(t) \\ y_{d_2}(t) \end{bmatrix}. \tag{8.47}$$

For the purpose of simulation, let the initial state $x(0)$ be equal to $[5\ 15\ 10\ 15\ 10\ 15]$. The output and input of the system $\mathscr{S}$ are depicted under both of the controllers K_c and K_d in Figures 8.4 and 8.5, respectively. These figures demonstrate that the controller K_d is a satisfactory decentralized implementation of K_c, as it can generate trajectories very close to the desired ones produced by K_c. This is a consequence of the fact that K_d^1 captures the dynamics of the centralized closed-loop system, and the sub-controller K_d^2 is mainly required for the internal stability of the decentralized closed-loop system. Note that in order to make the decentralized trajectories closer to the centralized ones particularly in the time interval $[0, 50]$, one needs to design an observer-based low-level controller K_d^2 using the method developed here, rather than searching for the best static low-level controller.

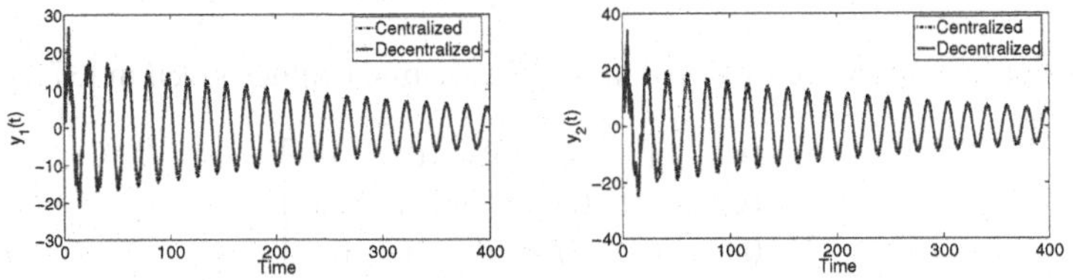

Fig. 8.4 The output of the system $\mathscr{S}$ under the controllers K_c and K_d.

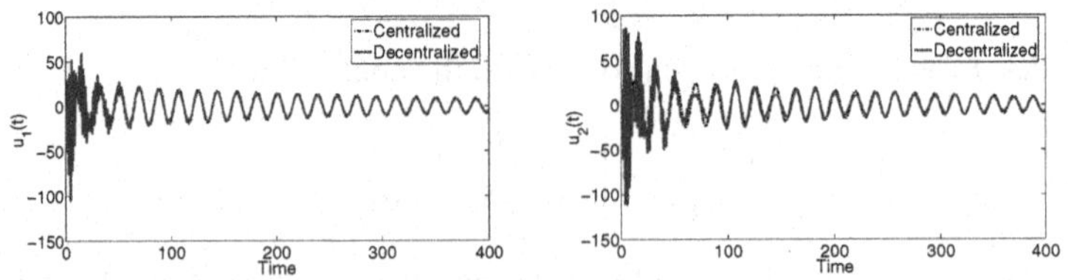

Fig. 8.5 The input of the system $\mathscr{S}$ under the controllers K_c and K_d.

8.5 Summary

This chapter is concerned with the decentralized implementation of centralized controllers for strongly connected interconnected systems. A parameterized decentralized controller is designed for a given centralized controller associated with an interconnected system. This two-level decentralized controller is composed of two interconnected decentralized sub-controllers, where the high-level sub-controller captures the dynamics of the system under the centralized controller. The low-level decentralized sub-controller is designed in such a way that the input and state trajectories of the system under the designed (overall) decentralized controller can become arbitrarily close to those of the system under the prescribed centralized controller by tuning the free parameters of the low-level sub-controller. This is carried out using the new notion of structural initial value observability. It is shown that the developed technique can shed light on some aspect of the LQR optimal decentralized control problem.

The present work shows that every generic centralized controller is decentrally implementable, but the order of the obtained decentralized controller is high, due to the structure of its high-level sub-controller. Since every local controller of the high-level sub-controller is stable, one can use a model reduction technique to first reduce the order of the high-level sub-controller and then design the low-level sub-controller. Note that the order of the designed controller being large is a common issue even for classical centralized control problems such as H_2 or H_∞ optimal control and strong stabilization problems.

8.6 Appendix 1

A real algebraic variety is defined to be the set of real zeros of a multivariate polynomial with a positive degree. A matrix $\mathscr{M} \in \mathbf{R}^{p\times q}$ is said to be generic with respect to a given real algebraic variety (with a positive codimension) if it does not belong to that variety. Note that almost all matrices in $\mathbf{R}^{p\times q}$ are generic with respect to a fixed real algebraic variety. For simplicity, the term "generic with respect to a certain real algebraic variety" will be abbreviated as "generic" throughout this chapter. The controller K_c is said to be generic if the parameter set (A_o, B_o, C_o, D_o) does not belong to a specific real algebraic variety.

To prove that Properties (i)-(vi) given in Theorem 2 hold if $F(A_o, B_o, C_o, D_o) \neq 0$ for some polynomial F, it suffices to show that each of these properties holds for a generic controller K. It will be later studied in Appendix 2 how to obtain the polynomial F.

8.6.1 Static centralized controllers

Assume for now that K_c is simply a static controller with the gain K (i.e., $u(t) = Ky(t)$). The results will be extended to a dynamic controller K_c in the next subsection. Notice that $\tilde{\mathscr{S}}$ can be represented as

$$\begin{aligned}\dot{\tilde{x}}(t) &= \tilde{A}\tilde{x}(t) + \sum_{i=1}^{\nu} \tilde{B}_i u_{d_i}(t) \\ y_{d_i}(t) &= \tilde{C}_i \tilde{x}(t), \quad \forall i \in \nu,\end{aligned} \tag{8.48}$$

where

$$\tilde{A} := \begin{bmatrix} A & B_1K_1C & B_2K_2C & \cdots & B_\nu K_\nu C \\ 0 & A+BKC & 0 & \cdots & 0 \\ 0 & 0 & A+BKC & \cdots & 0 \\ \vdots & \vdots & \vdots & \ddots & \vdots \\ 0 & 0 & 0 & \cdots & A+BKC \end{bmatrix},$$
$$\tilde{B} := \begin{bmatrix} 0 & 0 & \cdots & 0 \\ B & 0 & \cdots & 0 \\ 0 & B & \cdots & 0 \\ \vdots & \vdots & \ddots & \vdots \\ 0 & 0 & \cdots & B \end{bmatrix}, \quad \tilde{C} := \begin{bmatrix} C_1 & -C_1 & 0 & \cdots & 0 \\ C_2 & 0 & -C_2 & \cdots & 0 \\ \vdots & \vdots & \vdots & \ddots & \vdots \\ C_\nu & 0 & 0 & \cdots & -C_\nu \end{bmatrix}, \tag{8.49}$$

and $K_i \in \mathbf{R}^{m_i \times r}$, $\tilde{B}_i \in \mathbf{R}^{(\nu+1)n \times m}$ and $\tilde{C}_i \in \mathbf{R}^{r_i \times (\nu+1)n}$ are the ith block row, the ith block column and the ith block row of K, $\tilde{B}$ and $\tilde{C}$, respectively.

Proof of Part (i) of Theorem 2: The goal is to prove that $\tilde{\mathscr{S}}$ is generically strongly connected. To this end, let K be a matrix such that A and $A+BKC$ have disjoint eigenvalues and that

$$\begin{bmatrix} A-\lambda_j I & B \\ K_i C & 0 \end{bmatrix} = \text{full row rank}, \quad \forall i \in \nu,\ j \in \mathbf{n}. \tag{8.50}$$

Note that these properties hold for a generic K, due to Assumptions 1 and 3. Assume that $p, q \in \nu$, $p \neq q$, are two indices for which $C_p(sI-A)^{-1}B_q$ is not identically zero. It is desired to show that the transfer function $\tilde{C}_p(sI-\tilde{A})^{-1}\tilde{B}_q$ is nonzero as well. One can write

$$\tilde{C}_p(sI-\tilde{A})^{-1}\tilde{B}_q = C_p(sI-A)^{-1}B_qK_qC(sI-A-BKC)^{-1}B. \tag{8.51}$$

Since $C_p(sI-A)^{-1}B_q$ is not identically zero, there exists a mode λ_j such that this transfer function becomes infinity at $s=\lambda_j$. In light of the above equality, $\tilde{C}_p(sI-\tilde{A})^{-1}\tilde{B}_q$ is guaranteed to be nonzero if $K_qC(sI-A-BKC)^{-1}B$ is finite and has full row rank at $s=\lambda_j$. The finiteness of $K_qC(\lambda_jI-A-BKC)^{-1}B$ follows from the assumption that A and $A+BKC$ have disjoint eigenvalues. Regarding the rank of this quantity, one can use the LUD decomposition to obtain

$$K_qC(A+BKC-\lambda_jI)^{-1}B = \text{full rank} \iff \begin{bmatrix} A+BKC-\lambda_jI & B \\ K_qC & 0 \end{bmatrix} = \text{full rank}. \tag{8.52}$$

On the other hand, since the number of rows of K_qC is less than or equal to the number of columns of B, one can write

$$\begin{bmatrix} A+BKC-\lambda_jI & B \\ K_qC & 0 \end{bmatrix} = \text{full rank} \iff \begin{bmatrix} A-\lambda_jI & B \\ K_qC & 0 \end{bmatrix} = \text{full (row) rank}. \tag{8.53}$$

The right side of the above statement holds in light of the foregoing assumption. Thus, $\tilde{C}_p(sI-\tilde{A})^{-1}\tilde{B}_q$ is a nonzero transfer function. So far, it is shown that if there is an edge (q,p) in the structural graph of $\mathscr{S}$, the same edge must exist in the structural graph of $\tilde{\mathscr{S}}$ too. Since the structural graph of $\mathscr{S}$ is strongly connected (by Assumption 2), this property implies that the structural graph of $\tilde{\mathscr{S}}$ is strongly connected as well. ■

Proof of Part (ii) of Theorem 2: This part follows immediately from the upper block-triangular structure of the matrix $\tilde{A}$ given in (8.49). ■

Proof of Part (iii) of Theorem 2: The objective is to show that the system $\tilde{\mathscr{S}}$ is controllable for a generic controller K. To this end, let K be a matrix satisfying the two generic properties stated in the proof of Part (i) of Theorem 2. It is sufficient to prove that the matrix $[\tilde{A}-\sigma I\ \tilde{B}]$ is full rank for every eigenvalue σ of $\tilde{A}$. For this purpose, two cases can be considered as follows:

- *σ is equal to λ_i, for some $i \in \nu$:* To prove the underlying statement, it is enough to show that the null space of the matrix $[\tilde{A}-\sigma I\ \tilde{B}]$ is of dimension νm (recall that $\tilde{B} \in \mathbf{R}^{(\nu+1)n\times \nu m}$). Let $[\alpha_0^T\ \alpha_1^T\ \cdots\ \alpha_{2\nu}^T]^T$ be a vector in the null space of $[\tilde{A}-\sigma I\ \tilde{B}]$, where $\alpha_p \in \mathbf{R}^n$, $\forall p \in \{0\}\cup\nu$ and $\alpha_q \in \mathbf{R}^m$, $\forall q \in \{\nu+1,\cdots,2\nu\}$. One can write

$$(A-\sigma I)\alpha_0 + \sum_{j=1}^{\nu} B_jK_jC\alpha_j = 0, \tag{8.54a}$$

$$(A+BKC-\sigma I)\alpha_j + B\alpha_{j+\nu} = 0, \quad \forall j \in \nu. \tag{8.54b}$$

By assumption, σ is not an eigenvalue of $A+BKC$. Thus, the equation (8.54b) yields

$$\alpha_j = -(A+BKC-\sigma I)^{-1}B\alpha_{j+\nu}, \quad \forall j \in \nu. \tag{8.55}$$

The equations (8.54a) and (8.55) can be combined to deduce that

$$[\alpha_0^T\ \alpha_{\nu+1}^T\ \alpha_{\nu+2}^T\ \cdots\ \alpha_{2\nu}^T]^T \tag{8.56}$$

is in the null space of the matrix

$$\left[A-\sigma I\ \ -B_1K_1C(A+BKC-\sigma I)^{-1}B\ \cdots\ -B_\nu K_\nu C(A+BKC-\sigma I)^{-1}B\right]. \tag{8.57}$$

It follows from the argument made in the proof of Part (i) of Theorem 2 that the column space of the above matrix is identical to the column space of the matrix

$[A-\sigma I\ B_1\ \cdots\ B_\nu] = [A-\sigma I\ B]$ (because $K_jC(A+BKC-\sigma I)^{-1}B$ has full row rank, for every $j \in \nu$). Since the matrix $[A-\sigma I\ B]$ is full rank, it can be concluded that the null space of the matrix given in (8.57) is of dimension νm. As a result, the vector $[\alpha_0^T\ \alpha_{\nu+1}^T\ \alpha_{\nu+2}^T\ \cdots\ \alpha_{2\nu}^T]^T$ belongs to a νm-dimensional space. This observation and the equation (8.55) lead to the conclusion that the null space of the matrix $[\tilde{A}-\sigma I\ \tilde{B}]$ is of dimension νm. This completes the proof.

- *σ is equal to $\bar{\lambda}_i$, for some $i \in$ **n**:* Assume that there exists a nonzero vector $[\beta_0\ \beta_0\ \cdots\ \beta_\nu]$ such that $\beta_p \in \mathbf{R}^{1\times n}$, $\forall p \in \{0\} \cup \nu$, and

$$\left[\beta_0\ \beta_0 \cdots\ \beta_\nu\right]\left[\tilde{A}-\sigma I\ \tilde{B}\right] = 0. \tag{8.58}$$

This implies that

$$\beta_0(A-\sigma I) = 0 \tag{8.59a}$$

$$\beta_j\left[A+BKC-\sigma I\ B\right] + \beta_0\left[B_jK_jC\ 0\right] = 0, \quad \forall j \in \nu. \tag{8.59b}$$

On the other hand, σ is not an eigenvalue of A (by the assumption made earlier). This observation, together with the equation (8.59a), yields that $\beta_0 = 0$. Since the pair (A,B) is controllable, it can be concluded from the equation (8.59b) and the equality $\beta_0 = 0$ that $\beta_1 = \beta_2 = \cdots = \beta_\nu = 0$, which violates the assumption that $[\beta_0\ \beta_0\ \cdots\ \beta_\nu]$ is nonzero. ■

Lemma 1 *Given $i \in \{1,2,...,\nu-1\}$ and $j \in$ **n**, the matrix*

$$\begin{bmatrix} A-\bar{\lambda}_jI & B_1K_1C & B_2K_2C & \cdots & B_iK_iC \\ 0 & A+BKC-\bar{\lambda}_jI & 0 & \cdots & 0 \\ 0 & 0 & A+BKC-\bar{\lambda}_jI & \cdots & 0 \\ \vdots & \vdots & \vdots & \ddots & \vdots \\ 0 & 0 & 0 & \cdots & A+BKC-\bar{\lambda}_jI \\ C_1 & -C_1 & 0 & \cdots & 0 \\ C_2 & 0 & -C_2 & \cdots & 0 \\ \vdots & \vdots & \vdots & \ddots & \vdots \\ C_i & 0 & 0 & \cdots & -C_i \end{bmatrix} \tag{8.60}$$

is full rank for a generic controller K.

Proof: Let K be a matrix gain for which the following conditions are met:

- K is a block-diagonal matrix whose (l,l) block entry K_{ll} is of dimension $m_l \times r_l$, for every $l \in \nu$.
- The eigenvalues of $A+\sum_{l=1}^{i} B_lK_{ll}C_l$ and $A+BKC$ are disjoint .
- The pair $(A+BKC, C_l)$ is observable, for every $l \in \nu$.

Note that such a matrix K exists in light of the inequality $i < \nu$ together with Assumptions 1 and 2 (see [32]). Since the identity $B_lK_lC = B_lK_{ll}C_l$ holds for all $l \in \nu$, it is easy to show that the rank of the matrix in (8.60) is greater than or equal to the rank of the matrix in (8.61) given on top of the next page (note that all of the block entries of the latter matrix are equal to zero, except for the ones on the block diagonal and in the first block column). It follows from the aforementioned assumptions on the gain K that every block diagonal entry of the matrix given in (8.61) has full column rank. As a result, this matrix has full column rank as well, so does the matrix given in (8.60). So far, it is shown that the matrix (8.60) is full rank for a particular choice of K. Now, it is easy to verify that this implies that the matrix (8.60) must be full rank for a generic K. ■

$$\begin{bmatrix} \left[A+\sum_{l=1}^{i} B_l K_{ll} C_l - \bar{\lambda}_j I\right] & [0] & \cdots & [0] \\ \begin{bmatrix} A+BKC-\bar{\lambda}_j I \\ 0 \end{bmatrix} & \begin{bmatrix} A+BKC-\bar{\lambda}_j I \\ -C_1 \end{bmatrix} & \cdots & \begin{bmatrix} 0 \\ 0 \end{bmatrix} \\ \vdots & \vdots & \ddots & \vdots \\ \begin{bmatrix} A+BKC-\bar{\lambda}_j I \\ 0 \end{bmatrix} & \begin{bmatrix} 0 \\ 0 \end{bmatrix} & \cdots & \begin{bmatrix} A+BKC-\bar{\lambda}_j I \\ -C_i \end{bmatrix} \end{bmatrix} \tag{8.61}$$

Proof of Part (iv) of Theorem 2: The first objective is to show that the mode λ_j of the system $\tilde{\mathscr{S}}$ is observable, for every $j \in \mathbf{v}$. Let K be a matrix such that the sets of eigenvalues of $A+BKC$ and A are disjoint (note that this property holds for a generic K). Consider a vector $[\alpha_0^T \ \alpha_1^T \ \cdots \ \alpha_v^T]^T$ in the null space of $[(\tilde{A}-\lambda_j I)^T \ \tilde{C}^T]^T$, where $\alpha_i \in \mathbf{R}^n$, $\forall i \in \{0\} \cup \mathbf{v}$. It is intended to prove that this vector is equal to zero. To this end, one can write the following set of equations:

$$(A+BKC-\lambda_j I)\alpha_i = 0, \quad \forall i \in \mathbf{v}, \tag{8.62a}$$

$$C_i(\alpha_0 - \alpha_i) = 0, \quad \forall i \in \mathbf{v}, \tag{8.62b}$$

$$(A-\lambda_j I)\alpha_0 + \sum_{i=1}^{v} B_i K_i C \alpha_i = 0. \tag{8.62c}$$

The equation (8.62a) yields that $\alpha_1 = \alpha_2 = \cdots = \alpha_v = 0$. Now, it can be concluded from the equations (8.62b) and (8.62c) that $(A-\lambda_j I)\alpha_0 = C\alpha_0 = 0$. Since the pair (A,C) is observable, this implies that α_0 must be zero. As a result, λ_j is observable.

The second objective is to show that the repeated mode $\bar{\lambda}_j$ of the system $\tilde{\mathscr{S}}$ with multiplicity v is only $v-1$ times observable, for every $j \in \mathbf{n}$. For this purpose, let K be a matrix for which the three conditions given in the proof of Lemma 1 are satisfied (such as being block diagonal) and, in addition, the eigenvalues of $A+BKC$ are all distinct. Consider the matrix $[(\tilde{A}-\bar{\lambda}_j I)^T \ \tilde{C}^T]$. Replace the last block column of this matrix with the sum of all block columns of the matrix, remove its last block row and then re-arrange its block rows to obtain

$$\begin{bmatrix} [\Pi] & \begin{bmatrix} A+BKC-\bar{\lambda}_j I \\ \vdots \\ A+BKC-\bar{\lambda}_j I \end{bmatrix} \\ [0\,0\cdots 0] & [A+BKC-\bar{\lambda}_j I] \end{bmatrix}, \tag{8.63}$$

where Π is equal to the matrix given in (8.60) for $i=\nu-1$. The first observation is that the rank of the above matrix is less than or equal to the rank of the matrix $[(\tilde{A}-\bar{\lambda}_j I)^T\ \tilde{C}^T]$, due to the performed operations. Furthermore, the sub-matrix Π has full column rank in light of Lemma 1 and $\bar{\lambda}_j$ is a single eigenvalue of $A+BKC$ (by assumption). As a result, the rank of the matrix (8.63) is at least $\nu n-1$, and so is the rank of the matrix $[(\tilde{A}-\bar{\lambda}_j I)^T\ \tilde{C}^T]$. On the other hand, it is easy to verify that this observability matrix loses column rank (this can be seen by adding its block columns and checking the rank of the resulting block column), which makes its rank be at most $\nu n-1$. Hence, the rank of the matrix $[(\tilde{A}-\bar{\lambda}_j I)^T\ \tilde{C}^T]$ must be exactly $n\nu-1$, meaning that $\bar{\lambda}_j$ is exactly $\nu-1$ times observable for this choice of K. Let α_i denote a right eigenvector of $A+BKC$ associated with $\bar{\lambda}_i$, for every $i\in\mathbf{n}$. It is evident that the vector

$$\big[\underbrace{\alpha_i^T\ \alpha_i^T\ \cdots\ \alpha_i^T}_{\nu+1\text{ times}}\big]^T \tag{8.64}$$

is in the null space of $[(\tilde{A}-\bar{\lambda}_i I)^T\ \tilde{C}^T]$. Since the eigenvalues of $A+BKC$ are all distinct, this matrix has n independent eigenvectors; therefore, every vector in Φ can be written as a linear combination of the vectors in the form of (8.64) for $i\in\mathbf{n}$. This implies that Φ is in the unobservable subspace of $\tilde{\mathscr{S}}$. Due to the fact that this system has n unobservable modes and the dimension of Φ is exactly n, it can be concluded that the unobservable subspace of $\tilde{\mathscr{S}}$ is the same as Φ.

So far, it is proved that for a particular gain K, the repeated mode $\bar{\lambda}_j$ is only $\nu-1$ times observable and the unobservable subspace of the system is Φ. Now, it is straightforward to argue that these results are both valid for a generic controller K, on noting that:

- $\bar{\lambda}_j$ is at least one time unobservable for every arbitrary matrix K (this can be shown by adding up the block columns of the matrix $[(\tilde{A}-\bar{\lambda}_j I)^T\ \tilde{C}^T]$, as before).
- Since $\bar{\lambda}_j$ is only one time unobservable for a particular K, it must be at most one time unobservable for a generic K. ■

Proof of Part (v) of Theorem 2: The goal is to show that λ_j, $j\in\nu$, is not a DFM of the system $\tilde{\mathscr{S}}$. Since λ_j is a controllable and observable mode of the system $\tilde{\mathscr{S}}$ for a generic controller K (due to Parts (iii) and (iv) of the theorem), it is enough to prove that the inequality

$$\text{rank}\begin{bmatrix} \tilde{A}-\lambda_j I & \tilde{B}_{i_1} & \ldots & \tilde{B}_{i_p} \\ \tilde{C}_{i_{p+1}} & 0 & \ldots & 0 \\ \vdots & \vdots & \ddots & \vdots \\ \tilde{C}_{i_\nu} & 0 & \ldots & 0 \end{bmatrix} \geq n(\nu+1) \tag{8.65}$$

holds for every permutation $(i_1, i_2, ..., i_\nu)$ of the set $\{1, 2, ..., \nu\}$ and $p \in \{1, ..., \nu - 1\}$. The validity of the above relation will be shown only for the permutation $(i_1, i_2, ..., i_\nu) = (1, 2, ..., \nu)$, as the proof is similar for other permutations. For this purpose, consider a vector $[\alpha_0^T \ \alpha_1^T \ \cdots \alpha_{\nu+p}^T]^T$ in the null space of the matrix given in the left side of the inequality (8.65), where $\alpha_l \in \mathbf{R}^n$, $\forall l \in \{0\} \cup \nu$ and $\alpha_q \in \mathbf{R}^m$, $\forall q \in \{\nu+1, \cdots, \nu+p\}$. One can write

$$(A - \lambda_j I)\alpha_0 + \sum_{i=1}^{\nu} B_i K_i C \alpha_i = 0, \tag{8.66a}$$

$$(A + BKC - \lambda_j I)\alpha_i + B\alpha_{i+\nu} = 0, \quad \forall i \in \{1, 2, ..., p\}, \tag{8.66b}$$

$$(A + BKC - \lambda_j I)\alpha_i = 0, \quad \forall i \in \{p+1, p+2, ..., \nu\}, \tag{8.66c}$$

$$C_i(\alpha_0 - \alpha_i) = 0, \quad \forall i \in \{p+1, p+2, ..., \nu\}. \tag{8.66d}$$

Let K be a generic matrix satisfying the two properties that the eigenvalues of A and $A + BKC$ constitute disjoint sets and that the relation (8.50) holds. The equation (8.66c) yields

$$\alpha_{p+1} = \alpha_{p+2} = \cdots = \alpha_\nu = 0 \tag{8.67}$$

Furthermore, it follows from the equation (8.66b) that

$$\alpha_i = -(A + BKC - \lambda_j I)^{-1} B \alpha_{i+\nu}, \quad \forall i \in \{1, 2, ..., p\}. \tag{8.68}$$

Therefore, one can deduce from the equations (8.66), (8.67) and (8.68) that the vector $[\alpha_0^T \ \alpha_1^T \ \cdots \ \alpha_p^T]^T$ is in the null space of the matrix

$$\begin{bmatrix} A - \lambda_j I & -B_1 K_1 C (A + BKC - \lambda_j I)^{-1} B & \cdots & -B_p K_p C (A + BKC - \lambda_j I)^{-1} B \\ C_{p+1} & 0 & \cdots & 0 \\ \vdots & \vdots & \ddots & \vdots \\ C_\nu & 0 & \cdots & 0 \end{bmatrix} \tag{8.69}$$

On the other hand, since $K_l C (A + BKC - \lambda_j I)^{-1} B$ has full row rank for every $l \in \nu$ (see the proof of Part (i) of Theorem 2), the column space of the above matrix is identical to the column space of

$$\begin{bmatrix} A - \lambda_j I & B_1 & \cdots & B_p \\ C_{p+1} & 0 & \cdots & 0 \\ \vdots & \vdots & \ddots & \vdots \\ C_\nu & 0 & \cdots & 0 \end{bmatrix}. \tag{8.70}$$

Recall that the rank of this matrix is greater than or equal to n (by Assumption 1). This means that the rank of the matrix given in (8.69) is at least equal to n, which indicates that the dimension of its null space is at most pm. Thus, the vector $[\alpha_0^T \ \alpha_1^T \ \cdots \ \alpha_p^T]^T$ is contained in a pm-dimensional space. This fact, together with the relations (8.67) and (8.68), yields that the vector $[\alpha_0^T \ \alpha_1^T \ \cdots \alpha_{\nu+p}^T]^T$ is in a *pm*-dimensional space as well, which immediately proves the inequality (8.65).

Proof of Part (vi) of Theorem 2: It is shown in Part (iv) of the theorem that the mode $\bar{\lambda}_j$ with multiplicity ν is exactly $\nu - 1$ times observable for a generic controller K. This implies that $\bar{\lambda}_j$ is a DFM. That $\bar{\lambda}_j$ is a single DFM can be proven in line with the argument made in the proof of Part (v) of Theorem 2 (and using Lemma 1). ■

8.6.2 Extension to dynamic centralized controllers

To generalize the results of the previous subsection to a dynamic controller K_c, define

$$\bar{A} = \begin{bmatrix} A & 0_{n\times n_o} \\ B_o C & A_o \end{bmatrix}, \quad \bar{B} = \begin{bmatrix} B \\ 0_{n_o\times m} \end{bmatrix}, \quad \bar{C} = \begin{bmatrix} C & 0_{r\times n_o} \\ 0_{n_o\times n} & I_{n_o} \end{bmatrix}, \quad \bar{K} = \begin{bmatrix} D_o & C_o \end{bmatrix}. \tag{8.71}$$

It is easy to verify that the state-space matrices of the system $\tilde{\mathscr{S}}$ with the realization (8.48) are

$$\tilde{A} := \begin{bmatrix} A & B_1\bar{K}_1\bar{C} & B_2\bar{K}_2\bar{C} & \cdots & B_\nu\bar{K}_\nu\bar{C} \\ 0 & \bar{A}+\bar{B}\bar{K}\bar{C} & 0 & \cdots & 0 \\ 0 & 0 & \bar{A}+\bar{B}\bar{K}\bar{C} & \cdots & 0 \\ \vdots & \vdots & \vdots & \ddots & \vdots \\ 0 & 0 & 0 & \cdots & \bar{A}+\bar{B}\bar{K}\bar{C} \end{bmatrix} \tag{8.72}$$

and

$$\tilde{B} := \begin{bmatrix} 0 & 0 & \cdots & 0 \\ \bar{B} & 0 & \cdots & 0 \\ 0 & \bar{B} & \cdots & 0 \\ \vdots & \vdots & \ddots & \vdots \\ 0 & 0 & \cdots & \bar{B} \end{bmatrix}, \quad \tilde{C} := \begin{bmatrix} C_1 & -\bar{C}_1 & 0 & \cdots & 0 \\ C_2 & 0 & -\bar{C}_2 & \cdots & 0 \\ \vdots & \vdots & \vdots & \ddots & \vdots \\ C_\nu & 0 & 0 & \cdots & -\bar{C}_\nu \end{bmatrix}, \tag{8.73}$$

instead of the ones given in (8.49), where $\bar{K}_i \in \mathbf{R}^{m_i\times(r+n_o)}$ and $\bar{C}_i \in \mathbf{R}^{r_i\times(n+n_o)}$ are the ith block rows of $\bar{K}$ and $\bar{C}$, respectively, for every $i \in \nu$. It can be observed that $\tilde{A}$, $\tilde{B}$ and $\tilde{C}$ for a general controller K_c (given in (8.72) and (8.73)) resemble the corresponding matrices for a static controller K_c (given in (8.49)); more precisely:

- $\bar{A}$, $\bar{B}$, $\bar{C}$ and $\bar{K}$ used in (8.72) and (8.73) correspond to A, B, C and K used in (8.49).
- By fixing A_o and B_o, the parameters of the controller are C_o and D_o, which are combined in the matrix $\bar{K}$. In the static case, the parameter of the controller is K, i.e. the counterpart of $\bar{K}$.

Note that the above formulation is a well-known technique used in a number of papers, such as [13], to convert a dynamic decentralized controller problem to a static one. After recognizing this analogy between the static and dynamic cases, one can pursue the arguments made in the previous subsection to prove the stated results for a general controller K_c. For instance, the only modifications needed in the proofs of Parts (i) and (iii) of Theorem 2 are to replace the conditions

$$\begin{bmatrix} A-\lambda_j I & B \\ K_i C & 0 \end{bmatrix} = \text{full row rank}, \quad \forall i \in \nu,\ j \in \mathbf{n}, \text{ and } sp(A) \cap sp(A+BKC) = \text{empty} \tag{8.74}$$

with

$$\begin{bmatrix} \bar{A}-\lambda_j I & \bar{B} \\ \bar{K}_i \bar{C} & 0 \end{bmatrix} = \text{full row rank}, \quad \forall i \in \nu,\ j \in \mathbf{n}, \text{ and } sp(A) \cap sp(\bar{A}+\bar{B}\bar{K}\bar{C}) = \text{empty}, \tag{8.75}$$

where $sp(\cdot)$ is the spectral operator returning the set of eigenvalues of a matrix. Now, it suffices to notice that the new conditions hold for a generic controller K_c.

8.7 Appendix 2

This part aims to propose a method to find a nonzero polynomial F (whose coefficients are dependent only on the parameters of $\mathscr{S}$, rather than K_c) such that Properties (i)-(vi) stated in Theorem 2 hold whenever $F(A_o, B_o, C_o, D_o) \neq 0$. To this end, note that Property (ii) holds for all controllers K_c. Besides, Properties (v) and (vi) together imply Properties (iii) and (iv). Hence, it suffices to only consider Properties (i), (v) and (vi). The method proposed in [38] can be used for this purpose (because the underlying problem is a special case of the one tackled in [38]).

Given a controller K_c, notice that Properties (v) and (vi) hold if and only if there exists a block diagonal matrix $\tilde{K}$ (whose i^{th} block diagonal entry is of dimension $m_i \times r_i$, for every $i \in \nu$) such that $\det(sI-\tilde{A}-\tilde{B}\tilde{K}\tilde{C})\det(sI-\bar{A}-\bar{B}\bar{K}\bar{C})^{-1}$ and $\det(sI-A)\det(sI-\bar{A}-\bar{B}\bar{K}\bar{C})^{\nu-1}$ with the variable "s" have no common zero [13]. It can be concluded from the proof of Part (iv) of Theorem 2 that $\det(sI-\tilde{A}-\tilde{B}\tilde{K}\tilde{C})\det(sI-\bar{A}-\bar{B}\bar{K}\bar{C})^{-1}$ is a polynomial (as opposed to a non-polynomial rational function). In line with the proof of Theorem 1 in [38], define the polynomial $P(A_o, B_o, C_o, D_o, \tilde{K})$ as the determinant of the Sylvester matrix associated with the polynomials $\det(sI-\tilde{A}-\tilde{B}\tilde{K}\tilde{C})\det(sI-\bar{A}-\bar{B}\bar{K}\bar{C})^{-1}$ and $\det(sI-A)\det(sI-\bar{A}-\bar{B}\bar{K}\bar{C})^{\nu-1}$. Due to Sylvester's theorem, $P(A_o, B_o, C_o, D_o, \tilde{K})$ is nonzero if and only if the two polynomials mentioned above are co-prime (have no common zero). By fixing the terms A_o, B_o, C_o, D_o, the polynomial $P(A_o, B_o, C_o, D_o, \tilde{K})$ can be arranged in such a way that the monomials depend only on the entries of $\tilde{K}$ and the coefficients possibly depend on A_o, B_o, C_o and D_o. Since Properties (v) and (vi) hold for a generic K_c, it can be deduced that this resultant polynomial has at least one nonzero coefficient. Denoting this coefficient with $F_1(A_o, B_o, C_o, D_o)$, it follows from the argument made in [38] for a general scenario that Properties (v) and (vi) both hold if $F_1(A_o, B_o, C_o, D_o) \neq 0$. On the other hand, the proof of Part (i) of Theorem 2 and the discussion leading to (8.75) yield that Property (i) is guaranteed to hold if $\det(sI-A)$ and $\det(sI-\bar{A}-\bar{B}\bar{K}\bar{C})$ are co-prime, and in addition

$$\begin{bmatrix} \bar{A}-\lambda_j I & \bar{B} \\ \bar{K}_i\bar{C} & 0 \end{bmatrix} = \text{full row rank}, \quad \forall i \in \nu,\ j \in \mathbf{n}. \tag{8.76}$$

As before, it is easy to obtain a nonzero polynomial $F_2(A_o,B_o,C_o,D_o)$ such that these conditions all hold if $F_2(A_o,B_o,C_o,D_o) \neq 0$. Now, notice that the polynomial $F(A_o,B_o,C_o,D_o)$ can be taken as the least common multiple of $F_1(A_o,B_o,C_o,D_o)$ and $F_2(A_o,B_o,C_o,D_o)$.

References

1. S. Sojoudi and A. G. Aghdam, "Overlapping control systems with optimal information exchange," *Automatica*, vol. 45, no. 5, pp. 1176-1181, 2009.
2. A. I. Zecevic, G. Neskovic, and D. D. Siljak, "Robust decentralized exciter control with linear feedback," *IEEE Transactions on Power Systems*, vol. 19, no. 2, pp. 1096-1103, 2004.
3. R. Olfati-Saber, J. A. Fax and R. M. Murray, "Consensus and cooperation in networked multi-agent systems," *Proceedings of the IEEE*, vol. 95, no. 1, pp. 215-233, 2007.
4. J. Lavaei, A. Momeni, and A. G. Aghdam, "Spacecraft formation control in deep space with reduced communication requirement", *IEEE Transactions on Control System Technology*, vol. 16, no. 2, pp. 268-278, 2008.
5. K. L. Kosmatopoulos, E. B. Ioannou, and P. A. Ryaciotaki-Boussalis, "Large segmented telescopes: centralized decentralized and overlapping control designs," *IEEE Control Systems Magazine*, vol. 20, no. 5, pp. 59-72, 2000.
6. D. D. Siljak, Decentralized control of complex systems, Cambridge: Academic Press, 1991.
7. D. M. Stipanovic, G. Inalhan, R. Teo and C. J. Tomlin, "Decentralized overlapping control of a formation of unmanned aerial vehicles," *Automatica*, vol. 40 , no. 8, pp. 1285-1296, 2004.
8. A. Jadbabaie, J. Lin, and A. S. Morse, "Coordination of groups of mobile autonomous agents using nearest neighbor rules," *IEEE Transactions on Automatic Control*, vol. 48, no. 6, pp. 988-1001, 2003.
9. A. Kashyap, T. Basar and R. Srikant, "Quantized consensus," *Automatica*, vol. 43, no. 7, pp. 1192-1203, 2007.
10. M. Chiang, S. H. Low, A. R. Calderbank and J. C. Doyle, "Layering as optimization decomposition," *Proceedings of the IEEE*, vol. 95, no. 1, pp. 255-312, 2007.
11. S. H. Wang and E. J. Davison, "On the stabilization of decentralized control systems," *IEEE Transactions on Automatic Control*, vol. 18, no. 5, pp. 473-478, 1973.
12. B. L. O. Anderson, "Transfer function matrix description of decentralized fixed modes," *IEEE Transactions on Automatic Control*, vol. 27, no. 6, pp. 1176-1182, 1982.
13. E. J. Davison and T. N. Chang, "Decentralized stabilization and pole assignment for general proper systems," *IEEE Transactions on Automatic Control*, vol. 35, no. 6, pp. 652-664, 1990.
14. J. Lavaei and A. G. Aghdam, "A graph theoretic method to find decentralized fixed modes of LTI systems," *Automatica*, vol. 43, no. 12, pp. 2129-2133, 2007.
15. Z. Gong and M. Aldeen, "Stabilization of decentralized control systems," *Journal of Mathematical Systems, Estimation, and Control*, vol. 7, no. 1, pp. 1-16, 1997.
16. H. S. Witsenhausen, "A counterexample in stochastic optimum control," *SIAM Journal on Control and Optimization*, vol. 6, no. 1, pp. 131-147, 1968.
17. K. D. Young, "On near optimal decentralized control," *Automatica*, vol. 21, no. 5, pp. 607-610, 1985.
18. M. Rotkowitz and S. Lall, "A characterization of convex problems in decentralized control," *IEEE Transactions on Automatic Control*, vol. 51, no. 2, pp. 274-286, 2006.
19. J. Lavaei and A. G. Aghdam, "Decentralized control design for general interconnected systems with applications in cooperative control," *International Journal of Control*, vol. 80, no. 6, pp. 935-951, 2007.

20. V. Kariwala, "Fundamental limitation on achievable decentralized performance," *Automatica*, vol. 43, no. 10, pp. 1849-1854, 2007.
21. G. Zhai and M. Ikeda, "Decentralized H_∞ control of large-scale systems via output feedback," in *Proceedings of the 32nd Conference on Decision and Control*, San Antonio, TX, 1993.
22. R. J. Veillette, J. V. Medanić, and W. R. Perkins, "Design of reliable control systems," *IEEE Transactions on Automatic Control*, vol. 37, no. 3, pp. 290-304, 1992.
23. R. A. Paz and J. V. Medanić, "Decentralized H_∞ norm-bounding control for discrete-time systems," in *Proceedings of the 1992 American Control Conference*, Chicago, IL, 1992.
24. D. D. Sourlas and V. Manousiouthakis, "Best achievable decentralized performance," IEEE Transactions on Automatic Control, vol. 40, no. 11, pp. 1858-1871, 1995.
25. P. J. Campo and M. Morari, "Achievable closed-loop properties of systems under decentralized control: conditions involving the steady-state gain," vol. 39, no. 5, pp. 932-943, 1994.
26. G. C. Goodwin, M. E. Salgado, and E. I. Silva, "Time-domain performance limitations arising from decentralized architectures and their relationship to the RGA," *International Journal of Control*, vol. 78, no. 13, pp. 1045-1062, 2005.
27. H. Trinh and M. Aldeen, "A balancing realization approach to decentralized control of interconnected systems," in *Proceedings of the IEEE Singapore International Conference on Intelligent Control and Instrumentation*, Singapore, 1992.
28. G. C. Goodwin, M. M. Seron and M. E. Salgado, "H_2 design of decentralized controllers," in *Proceedings of the 1999 American Control Conference*, San Diego, CA, 1999.
29. R. A. Date and J. H. Chow, "A reliable coordinated decentralized control system design," in *Proceedings of the 28th Conference on Decision and Control*, Tampa, FL, 1989.
30. H. Kobayashi, H. Hanafusa, and T. Yoshikawa, "Controllability under decentralized information structure," *IEEE Transactions on Automatic Control*, vol. 23, no. 2, pp. 182-188, 1978.
31. B. D. O. Anderson and J. Moore, "Time-varying feedback laws for decentralized control," *IEEE Transactions on Automatic Control*, vol. 26, no. 5, pp. 1133-1139, 1981.
32. J. Lavaei and A. g. Aghdam, "Decentralized pole assignment for interconnected systems," in *Proceedings of the 2008 American Control conference*, Seattle, WA, 2008.
33. J. Lavaei and A. g. Aghdam, "Control of continuous-time LTI systems by means of structurally constrained controllers," *Automatica*, vol. 44, no. 1, pp. 141-148, 2008.
34. H. Kimura, "A new approach to the perfect regulation and the bounded peaking in linear multivariable control systems," *IEEE Transactions on Automatic Control*, vol. 26, no. 1, pp. 253-270, 1981.
35. J. Lavaei, and A. G. Aghdam, "Robust stability of LTI systems over semi-algebraic sets using sum-of-squares matrix polynomials," *IEEE Transactions on Automatic Control*, vol. 53, no. 1, pp. 417-423, 2008.
36. J. Lavaei and A. G. Aghdam, Performance improvement of robust controllers for polynomially uncertain systems, *Automatica*, vol. 46, no. 1, pp. 110-115, 2010.
37. A. G. Aghdam, E. J. Davison, and R. Becerril, "Structural modification of systems using discretization and generalized sampled-data hold functions," *Automatica*, vol. 42, no. 11, pp. 1935-1941, 2006.
38. S. Sojoudi, J. Lavaei, and A. G. Aghdam, "Robust stabilizability verification of polynomially uncertain LTI systems," *IEEE Transactions on Automatic Control*, vol. 52, no. 9, pp. 1721-1726, 2007.

Chapter 9
Robust Control of LTI Systems by Means of Structurally Constrained Controllers

9.1 Introduction

Numerous real-world systems can be envisaged as interconnected systems consisting of a number of subsystems [1]. Every controller for such a system is often composed of a set of local controllers corresponding to the individual subsystems. In an unconstrained control structure, each local controller has access to the outputs of all the subsystems. This class of controllers is referred to as centralized. However, in many control applications, each local controller can only use the information of a subset of subsystems. This control constraint is due, primarily, to some practical issues discussed below:

i) Interconnected systems often have several subsystems. Hence, a centralized controller for such large-scale systems can potentially be costly, in light of the required computations and transmission of information among the subsystems. In order to reduce the control expenditure for this type of systems, it is desirable to impose certain constraints on the control structure. A manifest example of this case is the traffic control system [2].
ii) For the interconnected systems with geographically distributed subsystems, transmission of information between two specific subsystems can be quite costly and prone to reliability problems. This is the case, for instance, in power systems, where two interacting power stations are located in remote places.
iii) In some interconnected systems, the output of certain subsystems may be inaccessible for some other subsystems in specific time intervals. This, for example, can occur frequently in the flight formation problem, due to the shadow phenomenon [3].

It follows from the above discussion that a structurally constrained controller is desirable for the interconnected systems [4]. The structure of such a controller is sometimes represented by a binary matrix, referred to as the information flow matrix [5]. Note that the information flow matrix corresponding to any system is part of the design specifications and is contingent upon the characteristics of the system and the control implementation cost as noted above.

A special case of structurally constrained controllers, often referred to as decentralized control, has been extensively studied in the literature [6, 7]. A decentralized controller comprises a number of non-interacting local controllers, which implies that the corresponding information flow matrix is block diagonal. Another type of structurally constrained control is the one in which some local controllers overlap in accordance with the overlapping structure of their corresponding subsystems [8, 9]. This class of control structure is called decentralized overlapping control, and is investigated in the literature in the Expansion-Inclusion framework [10].

The problem of stabilizability of systems (with known parameters) with respect to LTI *decentralized* and *decentralized overlapping* controllers has been investigated intensively, and several methodologies are presented accordingly for controller design [5, 11, 12, 13]. The notion of a decentralized fixed mode (DFM) was introduced in [6] to identify those modes of a LTI system (if any) which cannot be eliminated by means of a LTI decentralized controller. It is to be noted that the notion of DFMs for general proper systems is mainly characterized for block-diagonal information flow matrices. The notion of a decentralized overlapping fixed mode (DOFM) was introduced in [13] as a generalization of DFM, to identify those modes of a LTI system (if any) which are fixed with respect to the class of LTI structurally constrained controllers with any arbitrary information flow matrix. A simple graph-theoretic approach is also provided in [13] to obtain the DOFMs of any system efficiently.

The papers surveyed so far have merely considered the problem of structurally constrained stabilization for systems with known parameters. However, the real-world systems are uncertain to some degree. Under this circumstance, a region of uncertainty is usually envisaged to describe the range of uncertainty, and also some relations are considered to characterize the uncertain parameters of the system.

In the early works, the region of uncertainty was assumed to be the whole space and besides the uncertain parameters of the system were considered uncorrelated. The notions of structural controllability and structurally fixed mode were then defined based on these assumptions. Structural controllability was introduced in [14] to determine whether the uncontrollability of a LTI system is resulted from its structure or from the exact parameter matching in the system. Structural controllability is studied in several papers, e.g. see [15, 16, 17, 18]. Furthermore, the notion of a structurally fixed mode was defined in [19] to characterize those DFMs that are resulted from the structure of the system, and hence remain fixed regardless of the numerical values of the system's nonzero parameters.

Although the notions of structural controllability and structurally fixed modes are very useful in robust control design problems, they fail to address the very important practical issue of correlation between the nonzero parameters of the system [18]. In other words, in many practical problems different parameters of the system are correlated to each other, and belong to known regions in the parameter space. As a simple example, consider a RLC circuit and assume that the numerical values of its elements are known with a maximum of 10% error. In this case, every coefficient of the system transfer function can be written parametrically in terms of three quantities: the resistance, the capacitance and the inductance. This implies that all the coefficients of the transfer function are correlated and that the uncertainty region is indeed a cube. For instance, if one of the coefficients of the transfer function is equal to R (resistance) and another one is equal to $2R$, the variation of one coefficient is twice as large as the other one, while in the conventional robustness analysis their variations are considered independent of each other.

This chapter deals with the robust stabilizability of LTI systems via structurally constrained controllers. It is assumed that the state-space matrices of the system are polynomially uncertain, and that the uncertainty variables of the system belong to a known region. It is shown that if the system has no DOFMs at some point belonging to the region, then the points for which the system has a DOFM lie on an algebraic variety. As a result, if a system has no DOFM at a given nominal point, it almost always has no DOFMs at any operating point. Furthermore, since finding the exact algebraic variety can be formidable in general, a simple method is proposed to compute a dominant subset of it, in the sense that the dimension of this subset is greater than that of its complement. The two immediate applications of this work are encapsulated below:

- To control an uncertain interconnected system with unknown, nevertheless fixed or slowly varying parameters by means of a structurally constrained LTI controller, the prevailing method is to construct an adaptive control law [20, 21, 22]. Nevertheless, any adaptation law is feasible only if the system is stabilizable over the uncertainty region with respect to the underlying class of controllers. This chapter provides a systematic method to check this feasibility criterion. The results obtained can also be used to develop more effective adaptive laws, compared to the existing ones.
- Robust stability with respect to parameter variation is widely studied in the literature, which aims to discover whether a controller designed for a system in the nominal point can stabilize it over the whole region of uncertainty [23, 24]. However, these works are effective as long as the system is stabilizable over the uncertainty region. Consequently, this chapter provides a technique to verify structural stabilizability as a necessary condition for the robust stability of the system.

It is to be noted that the robust stabilizability problem has been investigated in a number of papers, e.g. see [25, 26, 27, 28, 29, 30, 31]. Nonetheless, these works formulate the problem in some special cases, e.g. SISO systems, centralized controllers or polytopic uncertainties. In contrast, this chapter tackles the robust stabilizability problem in the most general case. It also provides more general results compared to the ones presented in the literature for structural controllability and structurally fixed modes.

9.2 Robustness property of the modes of a LTI system

Consider an uncertain LTI interconnected system $\mathscr{S}(\alpha)$ with unknown, nevertheless fixed or slowly varying parameters consisting of ν subsystems with the following state-space representation:

$$\begin{aligned}
\dot{x}(t) &= A(\alpha)x(t) + \sum_{i=1}^{\nu} B_i(\alpha)u_i(t) \\
y_i(t) &= C_i(\alpha)x(t) + \sum_{j=1}^{\nu} D_{ij}(\alpha)u_j(t), \quad i \in \bar{\nu} := \{1, 2, ..., \nu\}
\end{aligned} \tag{9.1}$$

where:

- $x(t) \in \Re^n$ is the state, and $u_i(t) \in \Re^{m_i}$ and $y_i(t) \in \Re^{r_i}$, $i \in \bar{\nu}$, are the input and the output of the ith subsystem $S_i(\alpha)$, respectively.
- $\alpha = \begin{bmatrix} \alpha_1 \ \alpha_2 \ \cdots \ \alpha_\mu \end{bmatrix}$ represents the vector variable corresponding to the uncertainty in the system, and belongs to a given region $\mathscr{D}$ of dimension μ.
- $A(\alpha)$, $B_i(\alpha)$, $C_i(\alpha)$ and $D_{ij}(\alpha)$, $i, j \in \bar{\nu}$, are matrix polynomials of the variable α.

Define now:

$$\begin{aligned}
B(\alpha) &:= \begin{bmatrix} B_1(\alpha) \ B_2(\alpha) \ \cdots \ B_\nu(\alpha) \end{bmatrix}, \\
C(\alpha) &:= \begin{bmatrix} C_1(\alpha)^T \ C_2(\alpha)^T \ \cdots \ C_\nu(\alpha)^T \end{bmatrix}^T, \\
D(\alpha) &:= \begin{bmatrix} D_{11}(\alpha) & \cdots & D_{1\nu}(\alpha) \\ \vdots & \ddots & \vdots \\ D_{\nu 1}(\alpha) & \cdots & D_{\nu\nu}(\alpha) \end{bmatrix}
\end{aligned} \tag{9.2}$$

Assume that the system $\mathscr{S}(\alpha)$ is to be controlled by means of a structurally constrained controller. The constraint on the control structure determines which outputs y_i ($i \in \bar{\nu}$) are available to construct any specific input u_j ($j \in \bar{\nu}$) of the system. In order to simplify the formulation of the control constraint, a block matrix $\mathscr{K}$ with binary entries is defined, where its (i, j) block entry, $i, j \in \bar{\nu}$, is a $m_i \times r_j$ matrix with all elements equal to 1 if the output of the jth subsystem can contribute to the construction of the input of the ith subsystem, and is a $m_i \times r_j$ zero matrix otherwise. The matrix $\mathscr{K}$ represents the control constraint, and is referred to as the information flow matrix [5]. In the special case, when the entries of the matrix $\mathscr{K}$ are all equal to 1, the corresponding controller is centralized, and when $\mathscr{K}$ is block diagonal, the corresponding controller is decentralized. Throughout this chapter, the term *decentralized controller* is referred to the set of local controllers for an interconnected system with block-diagonal information flow matrix.

Consider the system $\mathscr{S}(\alpha)$ for an arbitrary value of α, namely α_0. The notion of a decentralized fixed mode (DFM) introduced in [5] for general proper systems corresponds to the modes of the system $\mathscr{S}(\alpha_0)$ which are fixed w.r.t. a *block-diagonal* information flow matrix $\mathscr{K}$. Decentralized overlapping fixed modes (DOFM) was defined in [13] to identify those modes of the system $\mathscr{S}(\alpha_0)$ which are fixed w.r.t. the LTI controllers complying with any arbitrary information flow matrix $\mathscr{K}$. It is to be noted that the term *decentralized overlapping control* is sometimes used in the literature for the particular case, when the information flow structure has the same overlapping structure as the subsystems of the system. However, the most general case of non-identical overlapping structure for the controller and the system will be considered in the present work.

Let **K** denote the space of all constant matrices of dimension $(m_1+\cdots+m_\nu)\times(r_1+\cdots+r_\nu)$ with the property that their zero entries coincide with those of the matrix $\mathscr{K}$. In fact, the space **K** parameterizes all the structurally constrained *static* controllers complying with the information flow matrix $\mathscr{K}$.

Lemma 1 *Given $\alpha_0\in\mathscr{D}$, assume that σ is a mode of the system $\mathscr{S}(\alpha_0)$. There exists no LTI structurally constrained controller complying with $\mathscr{K}$ to move the mode σ if and only if the following relation holds:*

$$\sigma\in sp\left\{A(\alpha_0)+B(\alpha_0)K(I-D(\alpha_0)K)^{-1}C(\alpha_0)\right\},\quad \forall K\in\mathbf{K} \tag{9.3}$$

Proof: For the case when $\mathscr{K}$ is block diagonal, the proof follows from the result obtained in [5] stating that if a mode is fixed w.r.t. any LTI *static* decentralized controller, it is then fixed w.r.t. all LTI dynamic decentralized controllers. The proof can be extended to the general case of non-block diagonal information flow structure, by noting that there exists a bijective morphism between DFMs and DOFMs (as substantiated in [13]). ■

The notion of a *structurally robust fixed mode* (SRFM) will now be introduced. The uncertain system $\mathscr{S}(\alpha)$ is said to have no SRFM in the region $\mathscr{D}$ w.r.t. $\mathscr{K}$, if there exists an α belonging to $\mathscr{D}$, denoted by α^*, such that the system $\mathscr{S}(\alpha^*)$ does not have any DOFM w.r.t. the information flow matrix $\mathscr{K}$. Note that if $\mathscr{S}(\alpha)$ has some SRFMs, then for any α_0 belonging to the region $\mathscr{D}$, the system $\mathscr{S}(\alpha_0)$ has at least one DOFM w.r.t. $\mathscr{K}$.

The following definitions will prove to be convenient in presenting the main results of this chapter.

Definition 1 *An algebraic variety refers to the set of common zeros of a number of polynomials. The notation $\mathscr{V}(f_1(\alpha),...,f_\lambda(\alpha))$ will be used throughout this chapter to refer to the algebraic variety generated by the common roots of the polynomials $f_1(\alpha),...,f_\lambda(\alpha)$.*

Definition 2 *An irreducible algebraic variety is said to be an affine variety.*

It is to be noted that in this chapter, only varieties in the real space (as opposed to the complex space) are considered. Hence, the term "real space" will be eliminated hereafter for simplicity. It is worth mentioning that an algebraic variety generated by a set of μ-variate polynomials is of dimension $\mu-1$. Nonetheless, this variety can be considered as the union of a number of affine varieties such that some of them are of pure dimension $\mu-1$ and the others have smaller dimensions.

Notation 1 *Given the variables $\alpha_1,\alpha_2,...,\alpha_\lambda$, all the monomials of degree ρ in the form of $\alpha_1^{\rho_1}\alpha_2^{\rho_2}...\alpha_\lambda^{\rho_\lambda}$ are denoted (in an arbitrary order) by $\Phi_\rho^1(\alpha_1,...,\alpha_\lambda)$, $\Phi_\rho^2(\alpha_1,...,\alpha_\lambda),...,\Phi_\rho^\xi(\alpha_1,...,\alpha_\lambda)$, where $\xi=\binom{\rho+\lambda-1}{\lambda-1}$.*

The next theorem presents the main result of this chapter.

Theorem 1 *The following statements are true for the robust stability of the system $\mathscr{S}(\alpha)$ in the region $\mathscr{D}$:*

a) Assume that the system $\mathscr{S}(\alpha^*)$ has no DOFM w.r.t. $\mathscr{K}$, for some $\alpha^* \in \mathscr{D}$. There exists an algebraic variety of dimension $\mu - 1$ such that for any arbitrary α_0 belonging to $\mathscr{D}$, the system $\mathscr{S}(\alpha_0)$ does not have any DOFM w.r.t. $\mathscr{K}$ if and only if α_0 does not pertain to this variety.
b) If there exists a point $\alpha^* \in \mathscr{D}$ such that the system $\mathscr{S}(\alpha^*)$ has no DOFMs w.r.t. $\mathscr{K}$, then for almost all values of α_0 belonging to $\mathscr{D}$ the system $\mathscr{S}(\alpha_0)$ has no DOFMs w.r.t. $\mathscr{K}$ either.
c) If the system $\mathscr{S}(\alpha)$ has at least a SRFM over the region $\mathscr{D}$ w.r.t. $\mathscr{K}$, then it also has some SRFMs over the whole space w.r.t. $\mathscr{K}$.

Proof of part (a): Define the following:

$$q(s,\alpha,K) = \det\left(sI - A(\alpha) - B(\alpha)K(I - D(\alpha)K)^{-1}C(\alpha)\right)\det(I - D(\alpha)K) \quad (9.4)$$

It is straightforward to show that $q(s,\alpha,K)$ is a polynomial in terms of the variables s, α and K, which are a scalar, a vector and a matrix, respectively. It follows from Lemma 1 that the system $\mathscr{S}(\alpha_0)$ has no DOFM w.r.t. $\mathscr{D}$ if and only if there exists a matrix $K_0 \in \mathbf{K}$ such that the polynomials $q(s,\alpha_0,K_0)$ and $q(s,\alpha_0,\bar{0})$ (which are functions of s only) are coprime, where $\bar{0}$ denotes a zero matrix with the same size as K_0. Note that the zeros of the polynomial $q(s,\alpha_0,\bar{0})$ are indeed the modes of the system $\mathscr{S}(\alpha_0)$. The coprimeness of these two polynomials will be formulated next.

One can rewrite the polynomial $q(s,\alpha,K)$ as:

$$q(s,\alpha,K) = \sum_{i=0}^{n} q_i(\alpha,K)s^i \quad (9.5)$$

for some polynomials $q_0(\alpha,K),\ldots,q_n(\alpha,K)$. Construct a $2n \times 2n$ Sylvester matrix by using the following rule:
Consider the first row of this matrix as:

$$\left[q_n(\alpha,K)\ q_{n-1}(\alpha,K)\ \cdots\ q_0(\alpha,K)\ 0\cdots\ 0\right] \quad (9.6)$$

and the $(n+1)^{th}$ row as:

$$\left[q_n(\alpha,\bar{0})\ q_{n-1}(\alpha,\bar{0})\ \cdots\ q_0(\alpha,\bar{0})\ 0\cdots\ 0\right] \quad (9.7)$$

Now, for any $i \in \{2,3,\ldots,n,n+2,\ldots,2n\}$, the ith row of the Sylvester matrix is obtained from the $(i-1)^{th}$ row by shifting it by one to the right and circularly shifting the rightmost entry to the leftmost position.
Denote the determinant of the resultant Sylvester matrix by the polynomial $r(\alpha,K)$. It can be inferred from the Sylvester's theorem that the polynomials $q(s,\alpha_0,K_0)$ and $q(s,\alpha_0,\bar{0})$ are coprime if and only if $r(\alpha_0,K_0) \neq 0$. One can conclude from this result and the existence condition given earlier for DOFMs of the system $\mathscr{S}(\alpha_0)$, that the system $\mathscr{S}(\alpha_0)$ has no DOFMs w.r.t. $\mathscr{K}$ if and only if there exists a matrix

$K_0 \in \mathbf{K}$ such that the polynomial $r(\alpha_0, K_0)$ is nonzero. This condition will be further simplified next.

Assume that the matrix $\mathscr{K}$ has j nonzero entries. Denote the nonzero values of the matrix $K \in \mathbf{K}$ with $k_1, k_2, ..., k_j$ in an arbitrary order. One can decompose the polynomial $r(\alpha, K)$ as follows:

$$r(\alpha, K) = \sum_{i=0}^{\bar{z}} r_i(\alpha) \Phi_z^i(k_1, k_2, ..., k_j) \tag{9.8}$$

for some polynomials $r_1(\alpha), ..., r_{\bar{z}}(\alpha)$, where $z = 2n(m_1 + \cdots + m_v)$ and $\bar{z} = \binom{z+j-1}{j-1}$. Note that the polynomials $r_1(\alpha), ..., r_{\bar{z}}(\alpha)$ cannot be all identical to zero. This results from the assumption that the system $\mathscr{S}(\alpha^*)$ has no DOFMs over the region $\mathscr{D}$ w.r.t. $\mathscr{K}$, which implies that $r(\alpha^*, K^*) \neq 0$ for some $K^* \in \mathbf{K}$. On the other hand, it can be easily verified that for a given scalar α_0, there exists a matrix $K_0 \in \mathbf{K}$ satisfying the relation $r(\alpha_0, K_0) \neq 0$ if and only if the polynomials $r_1(\alpha), ..., r_{\bar{z}}(\alpha)$ are not concurrently equal to zero at $\alpha = \alpha_0$.

The aforementioned results can be summarized as follows: *The system $\mathscr{S}(\alpha_0)$ has no DOFMs w.r.t. $\mathscr{K}$ if and only if α_0 does not pertain to the algebraic variety $\mathscr{V}(r_1(\alpha), ..., r_{\bar{z}}(\alpha))$*. This completes the proof of part (a).

Proof of part (b): The proof follows from part (a) on noting that the dimension of the algebraic variety $\mathscr{V}(r_1(\alpha), ..., r_{\bar{z}}(\alpha))$ is $\mu - 1$, while that of the region $\mathscr{D}$ is known to be equal to μ.

Proof of part (c): A proof by contradiction will be used here. Assume that the system $\mathscr{S}(\alpha)$ has no SRFM over the whole space w.r.t. $\mathscr{K}$, while it has some SRFMs over the region $\mathscr{D}$. It can be concluded from part (a) that there exists an algebraic variety such that the system $\mathscr{S}(\alpha)$ has DOFMs only for the uncertainties over this variety. Since the dimension of this variety is $\mu - 1$ (and that of the region $\mathscr{D}$ is μ), the region $\mathscr{D}$ is not contained by this variety. As a result, there exist infinitely many points belonging to $\mathscr{D}$, which do not belong to this algebraic variety, and consequently, their corresponding systems do not have any DOFMs. This contradicts the initial assumption that the system $\mathscr{S}(\alpha)$ has some SRFMs over the region $\mathscr{D}$ w.r.t. $\mathscr{K}$. ■

Remark 1 *As a by-products of part (b) of Theorem 1, if the system $\mathscr{S}(\alpha)$ has no DOFMs at the nominal point, say $\alpha = \alpha^*$, then the system almost always does not have a DOFM at any operating point, say $\alpha = \alpha_0$.*

Theorem 1 states that there is an algebraic variety in the region $\mathscr{D}$ for which the system $\mathscr{S}$ has some DOFMs w.r.t. $\mathscr{K}$. Hence, the identification of this variety can be helpful to provide a precise insight into the robust stability of the system. This algebraic variety is characterized by the polynomials $r_1(\alpha), r_2(\alpha), ..., r_{\bar{z}}(\alpha)$. Since $\bar{z}$ is typically a very large number, finding all these polynomials and the geometric shape of their common zeros are quite cumbersome. It is desired now to present a simple algorithm to obtain a dominant subset of this algebraic variety.

It is known from the discussion given earlier, that the algebraic variety $\mathscr{V}(r_1(\alpha), ..., r_{\bar{z}}(\alpha))$ is the union of a number of affine varieties. Some of these affine varieties are of pure dimension $\mu - 1$ and the remaining ones are of dimensions less than $\mu - 1$. More precisely, the dimensions of the latter subvarieties normally do not exceed $\max(0, \mu - \bar{z})$ (rather than $\mu - 2$), and since $\bar{z}$ is typically much greater than μ, these subvarieties should not exist in general [32]. Hence, one can rightly come to the conclusion that the affine varieties of pure dimension $\mu - 1$ play the primary role in the robust stability of the system $\mathscr{S}$, and the effect of other affine varieties (if any) is negligible. The following theorem states how the dominant part of $\mathscr{V}(r_1(\alpha), \ldots, r_{\bar{z}}(\alpha))$ can be identified efficiently.

Theorem 2 *Let the generic matrices K_1 and K_2 be chosen from the space* **K**. *Compute first the function $q(s, \alpha, K_i)$ and subsequently $r(\alpha, K_i)$ in terms of it, for $i = 1, 2$. Denote the greatest common divisor (gcd) of $r(\alpha, K_1)$ and $r(\alpha, K_2)$ with $h(\alpha)$. The variety $\mathscr{V}(h(\alpha))$ is included in the variety $\mathscr{V}(r_1(\alpha), \ldots, r_{\bar{z}}(\alpha))$, and also contains all the affine varieties of $\mathscr{V}(r_1(\alpha), \ldots, r_{\bar{z}}(\alpha))$ with the pure dimension $\mu - 1$.*

Proof: Define the following polynomial:

$$l(\alpha) = \gcd(r_1(\alpha), r_2(\alpha), \ldots, r_{\bar{z}}(\alpha)) \tag{9.9}$$

In light of the celebrated results in the area of algebraic sets, one can conclude that the variety $\mathscr{V}(l(\alpha))$ is a subset of the variety $\mathscr{V}(r_1(\alpha), \ldots, r_{\bar{z}}(\alpha))$, and moreover every affine variety in $\mathscr{V}(r_1(\alpha), \ldots, r_{\bar{z}}(\alpha))$ with the pure dimension $\mu - 1$ is a subset of $\mathscr{V}(l(\alpha))$ [32]. Therefore, to prove the theorem, it suffices to substantiate that $l(\alpha)$ is identical to $h(\alpha)$ for generic choices of K_1 and K_2. The latter statement can be proven easily, and the details are omitted here. ■

Theorem 2 proposes a simple method to obtain the dominant component of the variety, which broadly speaking, causes the uncertain system $\mathscr{S}(\alpha)$ not to be stabilizable by means of LTI structurally constrained controllers. It is to be noted that the varieties $\mathscr{V}(r_1(\alpha), \ldots, r_{\bar{z}}(\alpha))$ and $\mathscr{V}(h(\alpha))$ are the same in the univariate case ($\mu = 1$), but not necessarily in the multivariate case. This results from the fact that a set of multivariate polynomials can be relatively prime, while they have some common roots.

Remark 2 *The notion of structurally fixed modes introduced in [19] characterizes the DFMs of a system which remain fixed under any arbitrary perturbation of the nonzero parameters of the system. For a deterministic system $\mathscr{S}$, assume that the system matrices A, B, C and D have accumulatively e nonzero entries, and construct the uncertain replica of $\mathscr{S}$, denoted by $S(\alpha)$, and the region $\mathscr{D}$ as follows:*

- *Let the region $\mathscr{D}$ be $\Re^e$.*
- *Define the vector α as $\begin{bmatrix} \alpha_1 & \alpha_2 & \cdots & \alpha_e \end{bmatrix}$, and replace each nonzero entry of the system matrices with one of the variables $\alpha_1, \alpha_2, \ldots, \alpha_e$ so that none of these variables is recurrent. Denote the new matrices with $A(\alpha), B(\alpha), C(\alpha)$ and $D(\alpha)$, and the corresponding system with $S(\alpha)$.*

It can be easily shown that the notion of a structurally robust fixed mode for the system $\mathscr{S}(\alpha)$ is the same as the notion of a structurally fixed mode for the system $\mathscr{S}$. This implies that the robust stability framework considered here comprises one *of the relevant well-known results in the literature. In this case, Theorem 1 leads to the famous result that a system with no structurally fixed modes has generically no DFMs.*

Remark 3 *Using a technique similar to the one exploited in Remark 2, it can be easily shown that the formulation presented in this chapter and the corresponding developments encompass the existing results on structural controllability. In this case, Theorem 1 can be interpreted as the celebrated result that if a system is structurally controllable, then almost all systems with the same structure are also controllable.*

9.3 Numerical example

Consider an uncertain third-order system $\mathscr{S}(\alpha)$ with three SISO subsystems, where $\alpha = \begin{bmatrix} \alpha_1 & \alpha_2 \end{bmatrix}$, and assume the following parameters for its state-space representation:

$$\begin{aligned} A(\alpha) &= \begin{bmatrix} \alpha_2 & 0 & \alpha_1^3 - \alpha_2^3 \\ 0 & \alpha_1 & \alpha_1 - \alpha_2 \\ 0 & \alpha_1 & 0 \end{bmatrix}, \quad B(\alpha) = \begin{bmatrix} \alpha_1 & 0 & 0 \\ \alpha_2^2 & 0 & 0 \\ 0 & \alpha_1 & \alpha_1 + \alpha_2 \end{bmatrix}, \\ C(\alpha) &= \begin{bmatrix} \alpha_2 & 0 & 0 \\ 0 & 0 & \alpha_1^3 + \alpha_2^3 \\ \alpha_1 & 0 & 0 \end{bmatrix}, \quad D(\alpha) = \begin{bmatrix} \alpha_1 & 0 & 0 \\ 0 & 0 & \alpha_1 \\ \alpha_1 & 0 & \alpha_2 \end{bmatrix} \end{aligned} \tag{9.10}$$

Define now two regions of uncertainty as follows:

$$\mathscr{D}_1 = \{\alpha \,:\, 1 \leq \alpha_2^2 - 2\alpha_1^2 \leq 2\}, \quad \mathscr{D}_2 = \{\alpha \,:\, 1 \leq \alpha_1^2 - 2\alpha_2^2 \leq 2\} \tag{9.11}$$

It is desired to check the robust stability of the system $\mathscr{S}$ over the regions $\mathscr{D}_1$ and $\mathscr{D}_2$. To this end, two different control structures are delineated below:

1. Let the information flow matrix $\mathscr{K}$ be:

$$\mathscr{K} = \begin{bmatrix} 1 & 0 & 1 \\ 0 & 0 & 0 \\ 0 & 0 & 0 \end{bmatrix} \tag{9.12}$$

   In light of Theorem 2, two generic matrices should be chosen first. Let these matrices be:

$$K_1 = \begin{bmatrix} -1 & 0 & 6 \\ 0 & 0 & 0 \\ 0 & 0 & 0 \end{bmatrix}, \quad K_2 = \begin{bmatrix} 10 & 0 & 2 \\ 0 & 0 & 0 \\ 0 & 0 & 0 \end{bmatrix} \tag{9.13}$$

   The polynomials $r(s, K_1)$ and $r(s, K_2)$ can be simply obtained as:

$$\begin{aligned} r(s,K_1) &= -\alpha_2^4\alpha_1^3(-\alpha_1-\alpha_2+\alpha_1^2\alpha_2^2+\alpha_2^3\alpha_1+\alpha_2^4) \\ &\quad \times(\alpha_2^2+\alpha_1\alpha_2+\alpha_1^2)^2(6\alpha_1-\alpha_2)^3(\alpha_1-\alpha_2)^3, \\ r(s,K_2) &= -8\alpha_2^4\alpha_1^3(-\alpha_1-\alpha_2+\alpha_1^2\alpha_2^2+\alpha_2^3\alpha_1+\alpha_2^4) \\ &\quad \times(\alpha_2^2+\alpha_1\alpha_2+\alpha_1^2)^2(5\alpha_2+\alpha_1)^3(\alpha_1-\alpha_2)^3 \end{aligned} \tag{9.14}$$

Consequently:

$$\begin{aligned} h(\alpha) &= \gcd(r(s,K_1),r(s,K_2)) \\ &= \alpha_2^4\alpha_1^3(-\alpha_1-\alpha_2+\alpha_1^2\alpha_2^2+\alpha_2^3\alpha_1+\alpha_2^4)(\alpha_2^2+\alpha_1\alpha_2+\alpha_1^2)^2(\alpha_1-\alpha_2)^3 \end{aligned} \tag{9.15}$$

Since the real values of α are of importance, one can consider the variety $\mathscr{V}(p(\alpha))$, instated of $\mathscr{V}(h(\alpha))$, where:

$$p(\alpha) = (-\alpha_1-\alpha_2+\alpha_1^2\alpha_2^2+\alpha_2^3\alpha_1+\alpha_2^4)(\alpha_1-\alpha_2) \tag{9.16}$$

It is worth noting that the curve $\alpha_1-\alpha_2$ corresponds to the specific perturbations which make the system unobservable (and hence, the corresponding DOFMs are, in fact, centralized fixed modes as well), while the curve $-\alpha_1-\alpha_2+\alpha_1^2\alpha_2^2+\alpha_2^3\alpha_1+\alpha_2^4$ represents the perturbations which generally lead to observable and controllable fixed modes. One can verify that for this example, the exact algebraic variety $\mathscr{V}(r_1(\alpha),...,r_{\bar{z}}(\alpha))$ introduced in Theorem 1 is the same as $\mathscr{V}(p(\alpha))$ over the field of real numbers.

In order to verify whether or not the system $\mathscr{S}(\alpha)$ has some DOFMs over the region $\mathscr{D}_1$ w.r.t. $\mathscr{K}$, it suffices to check if the polynomial $p(\alpha)$ takes zero values in this region. For any point (α_1,α_2) belonging to $\mathscr{D}_1$, the inequalities $\alpha_2 > \sqrt{2}\alpha_1$ and $\alpha_2 > 1$ both hold. The first inequality implies that $\alpha_1-\alpha_2 \neq 0$. The second inequality, on the other hand, points to the fact that $-\alpha_1-\alpha_2+\alpha_1^2\alpha_2^2+\alpha_2^3\alpha_1+\alpha_2^4 > 0$. This means that $p(\alpha)$ cannot vanish in the region $\mathscr{D}_1$, or equivalently, that the system $\mathscr{S}(\alpha)$ has no DOFM over the region $\mathscr{D}_1$ w.r.t. $\mathscr{K}$. In contrast, it is easy to show that the polynomial $p(\alpha)$ is equal to zero for $(\alpha_1,\alpha_2)=(1.5312,0.75)$. Since this point belongs to the region $\mathscr{D}_2$, the system $\mathscr{S}(\alpha)$ has some DOFMs for certain points in the region $\mathscr{D}_2$ w.r.t. $\mathscr{K}$.

2. Assume now that the information flow matrix $\mathscr{K}$ is:

$$\mathscr{K} = \begin{bmatrix} 1 & 0 & 1 \\ 0 & 1 & 0 \\ 1 & 0 & 1 \end{bmatrix} \tag{9.17}$$

Pursuing the methodology given in the preceding case, it can be easily deduced that:

$$h(\alpha) = (\alpha_1-\alpha_2)^3 \tag{9.18}$$

Hence, the variety $\mathscr{V}(r_1(\alpha),...,r_{\bar{z}}(\alpha))$ includes a line and some affine varieties of dimensions less than $\mu-1=1$. This implies that any of these affine varieties should be of dimension 0, comprising a finite number of points. On the other hand, none of the regions $\mathscr{D}_1$ and $\mathscr{D}_2$ has an intersection with the line $\alpha_1 - \alpha_2 = 0$. This means that the system $\mathscr{S}$ has no DOFMs w.r.t. $\mathscr{K}$ in the uncertainty regions $\mathscr{D}_1$ and $\mathscr{D}_2$, except possibly for a finite number of points.

9.4 Summary

This chapter investigates the robust control of LTI systems subject to polynomial uncertainties over a given region. It is shown that if the system is stabilizable at some point in the given region by means of a structurally constrained controller, then it is also stabilizable via a controller of the same structure at any point in the region, as long as those points do not lie on an algebraic variety. This result shows that if the nominal model of the system is stabilizable by means of a structurally constrained controller, so is the system at almost all operating points. A numerical example is given to illustrate the importance of the results obtained and their efficacy in dealing with the most general forms of uncertainties. The proposed formulation and the corresponding developments provide a generalization of the existing results presented in the literature on the structural controllability and structurally fixed modes.

References

1. J. Lavaei and A. G. Aghdam, "Decentralized control design for interconnected systems based on a centralized reference controller," in *Proceedings of 45th IEEE Conference on Decision and Control*, San Diego, CA, 2006.
2. A. G. Aghdam, E. J. Davison, and R. B. Arreola, "Structural modification of systems using discretization and generalized sampled-data hold functions," *Automatica*, vol. 42 , no. 11, pp. 1935-1941, 2006.
3. R. S. Smith and F. Y. Hadaegh, "Control topologies for deep space formation flying spacecraft," in *Proceedings of 2002 American Control Conference*, Anchorage, AK, 2002.
4. J. Lavaei and A. G. Aghdam, "Optimal periodic feedback design for continuous-time LTI systems with constrained control structure," *International Journal of Control*, vol. 80, no. 2, pp. 220-230, 2007.
5. E. J. Davison and T. N. Chang, "Decentralized stabilization and pole assignment for general proper systems," *IEEE Transactions on Automatic Control*, vol. 35, no. 6, pp. 652-664, 1990.
6. S. H. Wang and E. J. Davison, "On the stabilization of decentralized control systems," *IEEE Transactions on Automatic Control*, vol. 18, no. 5, pp. 473-478, 1973.
7. J. Lavaei and A. G. Aghdam, "Simultaneous LQ control of a set of LTI systems using constrained generalized sampled-data hold functions," *Automatica*, vol. 43, no. 2, pp. 274-280, 2007.
8. D. D. Šiljak and A. I. Zecevic, "Control of large-scale systems: Beyond decentralized feedback," *Annual Reviews in Control*, vol. 29 , no.2, pp. 169-179, 2005.
9. A. I. Zecevic and D. D. Šiljak, "A new approach to control design with overlapping information structure constraints," *Automatica*, vol. 41, no. 2, pp. 265-272, 2005.
10. S. S. Stankovic, X. B. Chen and M. R. Matausek and D. D. Šiljak, "Stochastic inclusion principle applied to decentralized automatic generation control," *International Journal of Control*, vol. 72, no. 3, pp. 276-288, 1999.

11. J. Lavaei and A. G. Aghdam, "Elimination of fixed modes by means of high-performance constrained periodic control," in *Proceedings of 45th IEEE Conference on Decision and Control*, San Diego, CA, 2006.
12. Z. Gong and M. Aldeen, "Stabilization of decentralized control systems," *Journal of Mathematical Systems, Estimation, and Control*, vol. 7, no. 1, pp. 1-16, 1997.
13. J. Lavaei and A. G. Aghdam, "A necessary and sufficient condition for the existence of a LTI stabilizing decentralized overlapping controller," in *Proceedings of 45th IEEE Conference on Decision and Control*, San Diego, CA, 2006.
14. C. T. Lin, "Structural controllability," *IEEE Transactions on Automatic Control*, vol. 19, no. 3, pp. 201-208, 1974.
15. H. Mayeda, "On structural controllability theorem," *IEEE Transactions on Automatic Control*, vol. 26, no. 3, pp. 795-798, 1981.
16. S. Hosoe and K. Matsumoto, "On the irreducibility condition in the structural controllability theorem," *IEEE Transactions on Automatic Control*, vol. 24, no. 6, pp. 963-966, 1979.
17. K. J. Reinschke, F. Svaricek and H. D. Wend, "On strong structural controllability of linear systems," in *Proceedings of 31st IEEE Conference on Decision and Control*, 1992.
18. T. Yamada and D. Luenberger, "Generic controllability theorems for descriptor systems," *IEEE Transactions on Automatic Control*, vol. 30, no. 2, pp. 144-152, 1985.
19. M. E. Sezer and D. D. Šiljak, "Structurally fixed modes," *Systems & Control Letters*, vol. 1, no. 1, pp. 60-64, 1981.
20. K. A. Ossman, "Indirect adaptive control for interconnected systems," *IEEE Transactions on Automatic Control*, vol. 34, no. 8, pp. 908-911, 1989.
21. D. T. Gavel and D. D. Šiljak, "Decentralized adaptive control: structural conditions for stability," *IEEE Transactions on Automatic Control*, vol. 34, no. 4, pp. 413-426, 1989.
22. P. Ioannou and J. Sun, Robust adaptive control, Prentice Hall, 1996.
23. R. C.L.F Oliveira and P. L.D. Peres, "LMI conditions for robust stability analysis based on polynomially parameter-dependent Lyapunov functions," *Systems & Control Letters*, vol. 55, no. 1, pp. 52-61, 2006.
24. J. Lavaei and A. G. Aghdam, "A necessary and sufficient condition for robust stability of LTI discrete-time systems using sum-of-squares matrix polynomials," in *Proceedings of 45th IEEE Conference on Decision and Control*, San Diego, CA, 2006.
25. O. Bachelier, J. Bernussou, M. C. de Oliveira and J. C. Geromel, "Parameter dependant Lyapunov control design: numerical evaluation," in *Proceedings of 38th IEEE Conference on Decision and Control*, Phoenix, AZ, 1999.
26. D. Henrion, M. Sebek and V. Kucera, "Positive polynomials and robust stabilization with fixed-order controllers," *IEEE Transactions on Automatic Control*, vol. 48, no. 7, pp. 1178-1186, 2003.
27. A. Ghulchak, "Robust stabilizability and low-order controllers: duality principle in case studies," in *Proceedings of 43rd IEEE Conference on Decision and Control*, Atlantis, Bahamas, 2004.
28. M. Nikodem and A. W. Olbrot, "Robust stabilizability with phase perturbations," in *Proceedings of 34th IEEE Conference on Decision and Control*, 1995.
29. K. Benjelloun, E. K. Boukas and P. Shi, "Robust stabilizability of uncertain linear systems with Markovian jumping parameters," in *Proceedings of 1997 American Control Conference*, pp. 866-867, Albuquerque, New Mexico, 1997.
30. K. Wei and R. K. Yedavalli, "Robust stabilizability for linear systems with both parameter variation and unstructured uncertainty," *IEEE Transactions on Automatic Control*, vol. 34, no. 2, pp. 149-156, Feb. 1989.
31. H. Kwakernaak, "A condition for robust stabilizability," *Systems & Control Letters*, vol. 2, no. 1, Jul. 1982.
32. J. Harris, Algebraic geometry: a first course (graduate texts in mathematics), Springer-Verlag, 1992.

Chapter 10
Robust Control of Large-Scale Systems: Efficient Selection of Inputs and Outputs

10.1 Introduction

There has been a growing interest in recent years in robust control of systems with parametric uncertainty [1, 2, 3, 4, 5]. The dynamic behavior of this type of systems is typically governed by a set of differential equations whose coefficients belong to fairly-known uncertainty regions. Although there are several methods to capture the uncertain nature of a real-world system (e.g., by modeling it as a structured or unstructured uncertainty [6]), it turns out that the most realistic means of describing uncertainty is to parameterize it and then specify its domain of variation.

Robust stability is an important requirement in the control of a system with parametric uncertainty. This problem has been extensively studied in the case of linear time-invariant (LTI) control systems with specific types of uncertainty regions. for instance, sum-of-squares (SOS) relaxations are numerically efficient techniques introduced in [2] and [3] for checking the robust stability of polynomially uncertain systems. Moreover, a necessary and sufficient condition is proposed in [1] for the robust stability verification of this class of uncertain systems, by solving a hierarchy of semi-definite programming (SDP) problems.

The concepts of controllability and observability were introduced in the literature, and it was shown that they play a key role in various feedback control analysis and design problems such as model reduction, optimal control, state estimation, etc. [6]. Several techniques are provided in the literature to verify the controllability and/or observability of a system. However, in many applications it is important to know how much controllable or observable a system is. Gramian matrices were introduced to address this issue by providing a quantitative measure for controllability and observability [6]. While these notions were originally introduced for fixed known systems, they have been investigated thoroughly in the past two decades for the case of uncertain systems [7, 8, 9].

On the other hand, real-world systems are often composed of multiple interacting components, and hence possess sophisticated structures. Such systems are typically modeled as large-scale interconnected systems, for which classical control analysis and design techniques are usually inefficient. Several results are reported in the literature for structurally constrained control of large-scale systems in the contexts of decentralized and overlapping control, to address the shortcomings of the traditional control techniques [10, 11, 12, 13, 14].

This work aims to measure the minimum of the smallest singular value for the controllability and observability Gramians of parametric systems, over a given uncertainty region. Given a polynomially uncertain linear time-invariant (LTI) system with uncertain parameters defined on a semi-algebraic set, it is asserted that the controllability (observability) Gramian is a rational matrix in the corresponding parameters. It is desired to attain the minimum singular value of this matrix over the uncertainty region, but due to the rational structure of the matrix one cannot take advantage of the efficient techniques such as SOS tools. To bypass this obstacle, it is shown that said rational matrix can be replaced by a polynomial approximation which satisfies an important relation. An SOS formula is then obtained to find the underlying infimum. The special case of a polytopic uncertainty region is also investigated, due to its importance in practice. An alternative approach is proposed for this special case, with a substantially reduced computational burden.

Two primary applications of this work are as follows:

- To achieve robust closed-loop performance for a system subject to perturbation, it is very important to know the minimum energy required to control the system for any possible values of the parameters in the uncertainty region. This energy is known to be proportional to the inverse of the infimum of the singular values sought in this work. For instance, this infimum would determine if a system which is controllable for the nominal parameters, is also controllable with a sufficiently safe margin in a practical environment, where the parameters are subject to variation around the nominal values.
- In a real-world system with several interacting subsystems (which may be geographically distributed), it may not be feasible to establish information flow between all control agents. In other words, for such systems it is more desirable to have some form of decentralization, where each control input is constructed in terms of only those outputs which are available to the corresponding local controller due to the communication and computation limitations. To determine which inputs (or outputs) are most effective in the overall control operation and which ones are negligible, one can obtain the minimum input energy for different information flow structures, and choose the best structure by comparing the resultant values [15]. The results of the present work can be used to measure the required energy for different control structures systematically (this will be clarified in a numerical example).

10.2 Preliminaries and problem formulation

Consider an uncertain large-scale interconnected system with the following representation:

$$\begin{aligned} \dot{x}(t) &= A(\alpha)x(t) + B(\alpha)u(t) \\ y(t) &= C(\alpha)x(t) + D(\alpha)u(t) \end{aligned} \tag{10.1}$$

where

- $x(t) \in \mathbf{R}^n$, $u(t) \in \mathbf{R}^m$ and $y(t) \in \mathbf{R}^r$ are the state, input and output of the system, respectively.
- $\alpha := [\alpha_1, \alpha_2, \ldots, \alpha_k]$ denotes the vector of unknown, fixed uncertain parameters of the system.
- $A(\alpha), B(\alpha), C(\alpha)$ and $D(\alpha)$ are matrix polynomials in the variable α.

The system (10.1) is referred to as a *polynomially uncertain system.* Assume that the uncertainty vector α belongs to a given semi-algebraic set $\mathscr{D}$ characterized as follows:

$$\mathscr{D} = \{\alpha \in \mathbf{R}^k | f_1(\alpha) \geq 0, \ldots, f_z(\alpha) \geq 0\} \tag{10.2}$$

where $f_1(\alpha), \ldots, f_z(\alpha)$ are known scalar polynomials. Note that many practical uncertainty regions can be expressed either exactly or approximately in the above form.

Due to the large-scale nature of the system, the vectors $u(t)$ and $y(t)$ may contain several entries, and this has important implications in controller design, in general. Hence, assume that only a small part of the output vector, denoted by $\tilde{y}(t)$, and a small part of the input vector, denoted by $\tilde{u}(t)$, are desired to be used in the control structure, for the sake of simplifying the control operation. Let $\mathscr{S}(\alpha)$ be the system with the reduced input and output size, and denote its state-space representation as:

$$\begin{aligned} \dot{x}(t) &= A(\alpha)x(t) + \tilde{B}(\alpha)\tilde{u}(t) \\ \tilde{y}(t) &= \tilde{C}(\alpha)x(t) + \tilde{D}(\alpha)\tilde{u}(t) \end{aligned} \tag{10.3}$$

The objective is to evaluate the controllability/observability degradation resulted from reducing the size of the input and output of the system (10.1), i.e., it is intended to measure the controllability/observability degree of the system $\mathscr{S}(\alpha)$ in comparison to that of the original system (10.1). Addressing the above point is central to this chapter.

Suppose that the open-loop system $\mathscr{S}(\alpha)$ is robustly stable over the region $\mathscr{D}$, i.e., all eigenvalues of the matrix $A(\alpha)$ lie in the open left-half s-plane for every $\alpha \in \mathscr{D}$. This assumption is required for defining the infinite-horizon controllability Gramian of the system. It is noteworthy that the verification of the open-loop robust stability of the system $\mathscr{S}(\alpha)$ can be carried out systematically, using the existing methods in the literature such as the SOS technique given in [1] or the SDP method proposed in [3]. Denote the controllability and observability Gramians of the system $\mathscr{S}(\alpha)$ with $W_c(\alpha)$ and $W_o(\alpha)$, respectively, which are defined as follows:

$$W_c(\alpha) = \int_0^\infty e^{A(\alpha)t}\tilde{B}(\alpha)\tilde{B}(\alpha)^T e^{A(\alpha)^T t} dt \tag{10.4a}$$

$$W_o(\alpha) = \int_0^\infty e^{A(\alpha)^T t}\tilde{C}(\alpha)^T\tilde{C}(\alpha)e^{A(\alpha)t} dt \tag{10.4b}$$

In this chapter, the system $\mathscr{S}(\alpha)$ is said to be *robustly controllable* (*observable*) if it is controllable (observable) for all $\alpha \in \mathscr{D}$. Note that the robust controllability of the system $\mathscr{S}(\alpha)$ is equivalent to the positive-definiteness of the matrix $W_c(\alpha)$ for all $\alpha \in \mathscr{D}$. The Gramian matrix $W_c(\alpha)$ provides a measure for the degree of controllability (rather than just controllability/uncontrollability). Indeed, it is known that the input energy required for controlling the system is, roughly speaking, proportional to the inverse of the matrix $W_c(\alpha)$, and more specifically, proportional to the inverse of its smallest singular value (Proposition 4.5 in [6]). Hence, the robust controllability degree of the system $\mathscr{S}(\alpha)$ can be assessed in terms the minimum of the smallest singular value of the matrix $W_c(\alpha)$ over the region $\mathscr{D}$. This chapter is concerned with the computation of this minimum using an efficient method. Note that although this work focuses on the robust controllability problem, the results can be easily applied to the robust observability problem, as well (due to the existence of a natural duality between the two problems). Since the main development of this chapter is contingent upon the notion of sum-of-squares, some relevant background material is provided in the next subsection.

10.2.1 Background on Sum-of-Squares

This subsection aims to present important results in the area of sum-of-squares (SOS). A scalar polynomial $p(\alpha)$ is said to be SOS if there exists a set of scalar polynomials $p_1(\alpha), p_2(\alpha), ..., p_q(\alpha)$ such that:

$$p(\alpha) = p_1(\alpha)^2 + p_2(\alpha)^2 + \cdots + p_q(\alpha)^2, \quad \forall \alpha \in \mathbf{R}^k \tag{10.5}$$

It is evident that an SOS polynomial is always nonnegative, but the converse statement is not necessarily true. Indeed, a nonnegative polynomial cannot always be written as a sum of squared polynomials [16]. It is to be noted that checking whether a given polynomial is SOS amounts to solving a convex optimization problem [17]. Sparked by Polya's work, a great deal of effort has been made in the literature to make a connection between positive (nonnegative) polynomials and SOS polynomials. For instance, assume that $p(\alpha)$ is a homogeneous scalar polynomial which is strictly positive over the region $\mathbf{R}^k\backslash\{0\}$. Polya's theorem states that there exists a natural number ζ such that the coefficients of the polynomial $(\alpha_1 + \cdots + \alpha_k)^\zeta p(\alpha)$ are all nonnegative [18]. Hence, it is easy to show that $(\alpha_1^2 + \cdots + \alpha_k^2)^\zeta p(\alpha^2)$ is SOS, where $\alpha^2 := \begin{bmatrix} \alpha_1^2 & \cdots & \alpha_k^2 \end{bmatrix}$. Now, assume that $p(\alpha)$ is an arbitrary polynomial (not necessarily a homogeneous one). Obviously, if there exists a set of SOS polynomials $p_0(\alpha), p_1(\alpha), ..., p_k(\alpha)$ such that:

$$p(\alpha) = p_0(\alpha) + p_1(\alpha)f_1(\alpha) + \cdots + p_k(\alpha)f_k(\alpha) \tag{10.6}$$

then the polynomial $p(\alpha)$ is nonnegative over the region $\mathscr{D}$ (due to the fact that $p_0(\alpha), p_1(\alpha), ..., p_k(\alpha)$ are always nonnegative, and $f_1(\alpha), ..., f_k(\alpha)$ are also nonnegative over the region $\mathscr{D}$). The question arises as to whether the converse statement is true: given a polynomial $p(\alpha)$ that is nonnegative over the region $\mathscr{D}$, can it be written in the form (10.6) for some SOS polynomials $p_0(\alpha), p_1(\alpha), ..., p_k(\alpha)$? The answer to this question is negative. However, Putinar's theorem states that if the polynomial $p(\alpha)$ is strictly positive over the region $\mathscr{D}$, then it can be expressed as (10.6), provided the region $\mathscr{D}$ satisfies some mild conditions [19].

The above results will be used in the present chapter, but in a more general case when $p(\alpha)$ is a matrix polynomial.

10.3 Robust controllability degree

The following lemma is a generalized form of Lemma 1 in [3], and some interesting properties of the controllability Gramian.

Lemma 1 *There exist a matrix polynomial $H(\alpha)$ and a scalar polynomial $h(\alpha)$ such that the Gramian $W_c(\alpha)$ can be written as $\frac{H(\alpha)}{h(\alpha)}$, where:*

- *$H(\alpha)$ is positive-semidefinite for all $\alpha \in \mathscr{D}$.*
- *$h(\alpha)$ is strictly positive for all $\alpha \in \mathscr{D}$.*

Proof: It is known that the Gramian matrix $W_c(\alpha)$, $\alpha \in \mathscr{D}$, is the unique solution of the following continuous-time Lyapunov equation:

$$A(\alpha)W_c(\alpha) + W_c(\alpha)A(\alpha)^T = -\tilde{B}(\alpha)\tilde{B}(\alpha)^T \tag{10.7}$$

Hence, one can write:

$$[I \otimes A(\alpha) + A(\alpha) \otimes I] \operatorname{vec}\{W_c(\alpha)\} = \operatorname{vec}\left\{-\tilde{B}(\alpha)\tilde{B}(\alpha)^T\right\} \tag{10.8}$$

where $\otimes$ denotes the Kronecker product, and vec$\{\cdot\}$ is an operator which takes a matrix and converts it to a vector by stacking its columns on top of one another. Define now:

$$\begin{aligned} h(\alpha) &:= (-1)^n \det\{I \otimes A(\alpha) + A(\alpha) \otimes I\}, \quad \forall \alpha \in \mathbf{R}^k \\ H(\alpha) &:= h(\alpha)W_c(\alpha), \quad \forall \alpha \in \mathscr{D} \end{aligned} \tag{10.9}$$

Note that although $h(\alpha)$ is defined over the entire k-dimensional space, $H(\alpha)$ is defined only over the uncertainty region due to the fact that the Gramian $W_c(\alpha)$ used in its definition may not be well-defined outside the uncertainty region. It can be concluded from (10.8) and (10.9) that $H(\alpha)$ and $h(\alpha)$ are matrix and scalar polynomials, respectively, over the region $\mathscr{D}$. The domain of definition of $H(\alpha)$ can be easily extended to the whole space $\mathbf{R}^k$. Therefore, assume that $H(\alpha)$ and $h(\alpha)$ are two polynomials defined over the entire space, which satisfy the relations in (10.9) for every $\alpha \in \mathscr{D}$. To prove that $h(\alpha)$ is strictly positive over the uncertainty region, first define $\lambda_1(\alpha), \lambda_2(\alpha), ..., \lambda_n(\alpha)$ as the eigenvalues of the matrix $A(\alpha)$. Due to a well-known property of the Kronecker product, one can write:

$$h(\alpha) = (-1)^n \prod_{i=1}^{n} \prod_{j=1}^{n} (\lambda_i(\alpha) + \lambda_j(\alpha)) \tag{10.10}$$

Given a fixed $\alpha \in \mathscr{D}$, assume with no loss of generality that $\lambda_1(\alpha), ..., \lambda_q(\alpha)$ are real numbers and the remaining eigenvalues $\lambda_{q+1}(\alpha), ..., \lambda_n(\alpha)$ are non-real complex numbers. A few observations can be made as follows:

- For every $i, j \in \{1, 2, ..., n\}$ such that either $i > q$ or $j > q$, both of the terms $\lambda_i(\alpha) + \lambda_j(\alpha)$ and $\lambda_i(\alpha)^* + \lambda_j(\alpha)^*$ appear in the product given in the right side of the equation (10.10), which makes their product strictly positive.
- For every $i, j \in \{1, 2, ..., q\}$, the term $\lambda_i(\alpha) + \lambda_j(\alpha)$ is strictly negative (as the system is robustly stable). Moreover, there are q^2 of such terms in the equation (10.10).
- Since the real-valued matrix $A(\alpha)$ has an even number of non-real complex eigenvalues, $n - q$ is an even number.

These facts lead to the conclusion that the scalar polynomial $h(\alpha)$ is strictly positive for all $\alpha \in \mathscr{D}$. On the other hand, the definition of the Gramian $W_c(\alpha)$ implies that it is positive-semidefinite over the uncertainty region. As a result, the definition of $H(\alpha)$ in (10.9) yields that this matrix is positive-semidefinite for all $\alpha \in \mathscr{D}$. ■

In light of Lemma 1, the Gramian matrix $W_c(\alpha)$ is normally a non-polynomial rational function. This property impedes the use of the available SOS techniques, as they merely deal with polynomials. Hence, it is desirable to approximate $W_c(\alpha)$ by a polynomial. One naive way to do so is to replace the exponential terms in (10.4a) with some truncated Taylor series. However, the resultant polynomial approximation would not necessarily satisfy any important properties (namely the ones that will be introduced in Theorem 1 and are essential to the development of this chapter). Hence, a more advanced approximation technique will be provided in the sequel.

Notation 1 *Given a matrix M, $\underline{\sigma}\{M\}$ denotes its minimum singular value.*

Notation 2 *Given a vector $\beta = [\beta_1, \beta_2, \ldots, \beta_k]$, let β^2 be defined as $\beta^2 = [\beta_1^2, \beta_2^2, \ldots, \beta_k^2]$.*

Assumption 1 *The set $\mathscr{D}$ is compact, and there exist SOS scalar polynomials $w_0(\alpha), w_1(\alpha), ..., w_z(\alpha)$ such that all vectors α satisfying the inequality:*

$$w_0(\alpha) + w_1(\alpha) f_1(\alpha) + \cdots + w_z(\alpha) f_z(\alpha) \geq 0 \tag{10.11}$$

form a compact set.

There are two important points concerning Assumption 1. First, the validity of this assumption can be checked by solving a proper SOS problem. Furthermore, if the assumption does not hold, then the results of this chapter will become only sufficient, as opposed to both necessary and sufficient.

Consider $H(\alpha)$ and $h(\alpha)$ introduced in Lemma 1. It follows from the positiveness of $h(\alpha)$ and the compactness of $\mathscr{D}$ that there exist reals μ_1 and μ_2 such that:

$$0 < \mu_1 < h(\alpha) < \mu_2, \quad \forall \alpha \in \mathscr{D} \tag{10.12}$$

Definition 1 *Define $P_i(\alpha)$, $\alpha \in \mathscr{D}$, to be:*

$$P_i(\alpha) := W_c(\alpha) \times \left(1 - \left(1 - \frac{h(\alpha)}{\mu_2}\right)^{2i}\right), \quad i \in \mathbf{N} \tag{10.13}$$

The next theorem provides a means to approximate the rational controllability Gramian by a polynomial with any arbitrary precision.

Theorem 1 *The following statements are true:*

i) For every $i \in \mathbf{N}$, $P_i(\alpha)$ is a positive-semidefinite matrix polynomial over the region $\mathscr{D}$ that satisfies the matrix inequality:

$$A(\alpha)P_i(\alpha) + P_i(\alpha)A(\alpha)^T + \tilde{B}(\alpha)\tilde{B}(\alpha)^T \geq 0 \tag{10.14}$$

ii) Given $\alpha \in \mathscr{D}$, the sequence $\{P_i(\alpha)\}_1^\infty$ converges to $W_c(\alpha)$ from below monotonically, i.e. $P_1(\alpha) \leq P_2(\alpha) \leq P_3(\alpha) \leq \cdots$ and $\lim_{i\to\infty} \|P_i(\alpha) - W_c(\alpha)\| = 0$. Moreover:

$$P_i(\alpha) \leq W_c(\alpha) \leq \left(1 - \left(1 - \frac{\mu_1}{\mu_2}\right)^{2i}\right)^{-1} P_i(\alpha) \tag{10.15}$$

Proof of Part (i): The function $P_i(\alpha)$ being a matrix polynomial is a consequence of the following facts:

- In light of Lemma 1, $W_c(\alpha)$ can be written as $\frac{H(\alpha)}{h(\alpha)}$.
- The polynomial:

$$\left(1 - \left(1 - \frac{h(\alpha)}{\mu_2}\right)^{2i}\right) \tag{10.16}$$

  is divisible by $1 - (1 - \frac{h(\alpha)}{\mu_2})$, and hence is divisible by $h(\alpha)$ as well.

On the other hand, one can write:

$$\begin{aligned} A(\alpha)P_i(\alpha) + P_i(\alpha)A(\alpha)^T + \tilde{B}(\alpha)\tilde{B}(\alpha)^T &= \left(1 - \left(1 - \frac{h(\alpha)}{\mu_2}\right)^{2i}\right) \\ \times \left(A(\alpha)W_c(\alpha) + W_c(\alpha)A(\alpha)^T\right) + \tilde{B}(\alpha)\tilde{B}(\alpha)^T & \\ = \tilde{B}(\alpha)\tilde{B}(\alpha)^T \left(1 - \frac{h(\alpha)}{\mu_2}\right)^{2i} \geq 0 & \end{aligned} \tag{10.17}$$

This completes the proof.

Proof of Part (ii): The proof of this part follows directly from Definition 1 and the inequality (10.12). ■

Corollary 1 *The minimum singular values of $W_c(\alpha)$ and $P_i(\alpha)$ are related to each other by the following inequalities:*

$$\left(1-\left(1-\frac{\mu_1}{\mu_2}\right)^{2i}\right)\min_{\alpha\in\mathscr{D}}\underline{\sigma}\{W_c(\alpha)\}\leq\min_{\alpha\in\mathscr{D}}\underline{\sigma}\{P_i(\alpha)\}\leq\min_{\alpha\in\mathscr{D}}\underline{\sigma}\{W_c(\alpha)\} \quad (10.18)$$

where $i=1,2,\ldots$.

Proof: The proof is an immediate consequence of Theorem 1. ■

Theorem 1 shows that $W_c(\alpha)$ can be approximated, with any arbitrary precision, by a matrix polynomial satisfying a certain matrix inequality. Moreover, implicit bounds on the smallest singular value of the controllability matrix are provided in Corollary 1. In order to obtain a proper range of values for the degree of a polynomial which can approximate $W_c(\alpha)$ satisfactorily, it is required to know the ratio $\frac{\mu_1}{\mu_2}$ *a priori*. Fortunately, the relation (10.10) can be used to obtain this quantity, as it indicates that this ratio is, roughly speaking, related to the minimum and maximum eigenvalues of $A(\alpha)$ over the region $\mathscr{D}$. Note that the ratio $\frac{\mu_1}{\mu_2}$ quantifies the uncertainty degree of the open-loop system (i.e. the matrix $A(\alpha)$) in terms of the location of the eigenvalues. Let an optimization problem be introduced in the sequel.

Optimization 1 *Given the system $\mathscr{S}(\alpha)$ and the uncertainty region $\mathscr{D}$, maximize the real-valued scalar variable μ subject to the constraint that there exist a symmetric matrix polynomial $P(\alpha)$ and SOS matrix polynomials $S_0(\alpha),\ldots,S_z(\alpha),\tilde{S}_0(\alpha),\ldots,$ $\tilde{S}_z(\alpha)$ such that:*

$$A(\alpha)P(\alpha)+P(\alpha)A(\alpha)^T+\tilde{B}(\alpha)\tilde{B}(\alpha)^T=S_0(\alpha)+\sum_{i=1}^{z}S_i(\alpha)f_i(\alpha) \quad (10.19a)$$

$$P(\alpha)=\mu I_n+\tilde{S}_0(\alpha)+\sum_{i=1}^{z}\tilde{S}_i(\alpha)f_i(\alpha) \quad (10.19b)$$

where I_n is the $n\times n$ identity matrix. Denote the solution of this optimization problem with μ^*.

Theorem 2 *The quantity $\min_{\alpha\in\mathscr{D}}\underline{\sigma}\{W_c(\alpha)\}$ is equal to μ^*.*

Proof: Let $P(\alpha)$ be a matrix polynomial for which there exist a real μ and SOS matrix polynomials $S_0(\alpha),\ldots,S_z(\alpha),\tilde{S}_0(\alpha),\ldots,\tilde{S}_z(\alpha)$ satisfying the equalities given in (10.19). One can write:

$$A(\alpha)P(\alpha)+P(\alpha)A(\alpha)^T+\tilde{B}(\alpha)\tilde{B}(\alpha)^T\geq 0 \quad (10.20a)$$

$$P(\alpha)\geq\mu I_n \quad (10.20b)$$

for all $\alpha\in\mathscr{D}$. It follows from (10.7) and (10.20a) that:

$$A(\alpha)\left(W_c(\alpha)-P(\alpha)\right)+\left(W_c(\alpha)-P(\alpha)\right)A(\alpha)^T \leq 0, \quad \forall \alpha \in \mathscr{D} \tag{10.21}$$

Therefore, $P(\alpha) \leq W_c(\alpha)$, $\forall \alpha \in \mathscr{D}$. This, together with the inequality (10.20b), yields:

$$\mu \leq \min_{\alpha \in \mathscr{D}} \underline{\sigma}\{W_c(\alpha)\} \tag{10.22}$$

As a result:

$$\mu^* \leq \min_{\alpha \in \mathscr{D}} \underline{\sigma}\{W_c(\alpha)\} \tag{10.23}$$

On the other hand, notice that:

$$P_j(\alpha) > (\min_{\alpha \in \mathscr{D}} \underline{\sigma}\{P_j(\alpha)\} - \varepsilon)I_n, \quad \forall \alpha \in \mathscr{D} \tag{10.24}$$

for any $j \in \mathbf{N}$ and $\varepsilon \in \mathbf{R}^+$ (note that $\mathbf{R}^+$ represents the set of positive real numbers). Therefore, it can be inferred from Theorem 1 and Assumption 1 of the present work, and Theorem 2 in [21] (i.e. the general matrix form of Putinar's theorem given in the previous section) that there exist SOS matrix polynomials $\tilde{S}_0(\alpha), ..., \tilde{S}_z(\alpha)$ such that:

$$P_j(\alpha) = (\min_{\alpha \in \mathscr{D}} \underline{\sigma}\{P_j(\alpha)\} - \varepsilon)I_n + \tilde{S}_0(\alpha) + \sum_{i=1}^{z} \tilde{S}_i(\alpha) f_i(\alpha) \tag{10.25}$$

Now, recall from Theorem 1 that:

$$A(\alpha)P_j(\alpha) + P_j(\alpha)A(\alpha)^T + \tilde{B}(\alpha)\tilde{B}(\alpha)^T \geq 0 \tag{10.26}$$

Since the above inequality cannot be simply relaxed to strict inequality, Putnam's theorem cannot be used directly (as explained in Subsection 10.2.1). To bypass this obstacle, one can use the result of Theorem 2 in [21] that there exist SOS matrix polynomials $\bar{S}_0(\alpha), ..., \bar{S}_z(\alpha)$ such that:

$$\left(1 - \frac{h(\alpha)}{\mu_2}\right)^{2j} = \bar{S}_0(\alpha) + \sum_{i=1}^{z} \bar{S}_i(\alpha) f_i(\alpha) \tag{10.27}$$

because the left side of the above equation is always strictly positive over the region $\mathscr{D}$. Hence, it is concluded from (10.17) that:

$$\begin{aligned} A(\alpha)P_j(\alpha) + P_j(\alpha)A(\alpha)^T + \tilde{B}(\alpha)\tilde{B}(\alpha)^T &= \tilde{B}(\alpha)\tilde{B}(\alpha)^T \left(1 - \frac{h(\alpha)}{\mu_2}\right)^{2j} \\ &= \bar{S}_0(\alpha)\tilde{B}(\alpha)\tilde{B}(\alpha)^T \\ &\quad + \sum_{i=1}^{z} \bar{S}_i(\alpha)\tilde{B}(\alpha)\tilde{B}(\alpha)^T f_i(\alpha) \\ &= S_0(\alpha) + \sum_{i=1}^{z} S_i(\alpha) f_i(\alpha) \end{aligned} \tag{10.28}$$

where:

$$S_i(\alpha) := \bar{S}_i(\alpha)\tilde{B}(\alpha)\tilde{B}(\alpha)^T, \quad i = 0, 1, ..., k \tag{10.29}$$

are SOS matrix polynomials. It results from (10.25) and (10.28) that:

$$\min_{\alpha \in \mathscr{D}} \underline{\sigma}\{P_j(\alpha)\} - \varepsilon \leq \mu^* \tag{10.30}$$

The proof is completed by taking the inequalities (10.23) and (10.30) into consideration and letting i and ε go to infinity and zero, respectively, and by using Corollary 1. ∎

Theorem 2 provides a methodology for finding the quantity $\min_{\alpha \in \mathscr{D}} \underline{\sigma}\{W_c(\alpha)\}$ indirectly via solving Optimization 1. This optimization problem is in the form of SOS, which can be handled by a semi-definite program. For this purpose, one can use a proper software tool such as YALMIP or SOSTOOLS [22, 23]. Nevertheless, it is first required to consider some upper bounds *a priori* on the degrees of the polynomials involved in the corresponding optimization problem, from which a *lower bound* on the solution of Optimization 1 can be found. In other words, this optimization problem can be formulated as a hierarchy of SDP problems, whose solutions converge asymptotically to the quantity of interest, i.e. $\min_{\alpha \in \mathscr{D}} \underline{\sigma}\{W_c(\alpha)\}$, from below.

Remark 1 *Despite the fact that a high-order rational matrix ($W_c(\alpha)$ in the present problem) cannot, in general, be approximated satisfactorily by a low-order polynomial matrix $P(\alpha)$, it will be illustrated later in an example that a relatively low-order polynomial typically works well here. This is due to the fact that these two functions need to be sufficiently close only at a critical point corresponding to the solution of the optimization problem (as opposed to everywhere in the region $\mathscr{D}$).*

10.3.1 Special case: A polytopic region

Although Theorem 2 provides a numerically tractable method for measuring the robust controllability of a system, the proposed optimization problem can be simplified significantly for special cases of interest. For instance, assume that $\mathscr{D}$ is a polytopic region $\mathscr{P}$ given by:

$$\mathscr{P} = \{\alpha | \alpha_1 + \cdots + \alpha_k = 1, \ \alpha_1, ..., \alpha_k \geq 0\} \tag{10.31}$$

This type of uncertainty region is of particular interest, due to its important applications.

Assumption 2 *Assume that $A(\alpha)$ and $\tilde{B}(\alpha)$ are homogeneous matrix polynomials, and let their degrees be denoted by ζ_1 and ζ_2, respectively.*

Note that Assumption 2 holds automatically for polytopic systems, with $\zeta_1 = \zeta_2 = 1$.

Theorem 3 *The quantity* $\min_{\alpha\in\mathscr{P}} \underline{\sigma} W_c(\alpha)$ *is equal to the maximum value of* μ *for which there exists a homogeneous matrix polynomial* $\tilde{P}(\alpha)$ *satisfying the following inequalities for all* $\alpha \in \mathbf{R}^k$*:*

$$\tilde{P}(\alpha^2) \geq 0, \tag{10.32a}$$

$$\left[A(\alpha^2)\left(\tilde{P}(\alpha^2)+\mu(\alpha\alpha^T)^{\zeta_3}I_n\right)+\left(\tilde{P}(\alpha^2)+\mu(\alpha\alpha^T)^{\zeta_3}I_n\right)A^T(\alpha^2)\right]$$
$$\times(\alpha\alpha^T)^{\max(0,2\zeta_2-\zeta_1-\zeta_3)}+\tilde{B}(\alpha^2)\tilde{B}(\alpha^2)^T(\alpha\alpha^T)^{\max(0,\zeta_3+\zeta_1-2\zeta_2)}\geq 0 \tag{10.32b}$$

where ζ_3 *denotes the degree of the polynomial* $\tilde{P}(\alpha)$.

Proof: The proof follows by refining the proofs of Theorems 1 and 2 in such a way that the homogeneity of the system matrices as well as the polytopic structure of the uncertainty region are both taken into account. To this end, the homogenization technique given in [3] is adopted. First of all, notice that the equation (10.8) yields:

$$\mathrm{vec}\{W_c(\alpha)\} = -\left[I\otimes A(\alpha)+A(\alpha)\otimes I\right]^{-1}\mathrm{vec}\left\{\tilde{B}(\alpha)\tilde{B}(\alpha)^T\right\} \tag{10.33}$$

Note also that $I\otimes A(\alpha)+A(\alpha)\otimes I$ and $\tilde{B}(\alpha)\tilde{B}(\alpha)^T$ are both homogeneous polynomials. Therefore, one can conclude that the polynomials $H(\alpha)$ and $h(\alpha)$ introduced in Lemma 1 can both be assumed to be homogeneous. Now, let $P_i(\alpha)$ be defined as:

$$P_i(\alpha) := W_c(\alpha)\times\left(\left(\sum_{j=1}^{k}\alpha_j\right)^{2i\zeta_4}-\left(\left(\sum_{j=1}^{k}\alpha_j\right)^{\zeta_4}-\frac{h(\alpha)}{\mu_2}\right)^{2i}\right) \tag{10.34}$$

in lieu of the one defined in (10.13), where ζ_4 denotes the degree of $h(\alpha)$. Note that $P_i(\alpha)$ given above is identical to the one defined in (10.13) over the region $\mathscr{P}$, and its subtle difference is that it is homogeneous. In other words, $W_c(\alpha)$ is approximated by a homogeneous polynomial here. It is easy to show that Theorem 1 holds for the polynomial $P_i(\alpha)$ defined in (10.34). On the other hand, due to the polytopic nature of the set $\mathscr{P}$ in (10.31), the inequality (10.14) in Theorem 1 can be rewritten as:

$$\begin{aligned}&\left(A(\alpha)P_i(\alpha)+P_i(\alpha)A(\alpha)^T\right)\left(\sum_{i=j}^{k}\alpha_j\right)^{\max(0,2\zeta_2-\zeta_1-\zeta_3^i)}\\&+\tilde{B}(\alpha)\tilde{B}(\alpha)^T\left(\sum_{j=1}^{k}\alpha_j\right)^{\max(0,\zeta_1+\zeta_3^i-2\zeta_2)}\geq 0,\ \forall\alpha\in\mathscr{P}\end{aligned} \tag{10.35}$$

where ζ_3^i denotes the degree of $P_i(\alpha)$. One can write:

$$\begin{aligned}\tilde{P}_i(\alpha) &:= P_i(\alpha) - \min_{\alpha \in \mathscr{P}} \{P_i(\alpha)\} \left(\sum_{j=1}^{k} \alpha_j\right)^{\zeta_3^i} I_n \\ &= P_i(\alpha) - \min_{\alpha \in \mathscr{P}} \{P_i(\alpha)\} I_n \geq 0, \quad \forall \alpha \in \mathscr{P}\end{aligned} \tag{10.36}$$

Hence:

$$\tilde{P}_i(\alpha) \geq 0, \quad \forall \alpha \in \mathscr{P} \tag{10.37}$$

Notice now the following facts:

- A matrix polynomial $M(\alpha)$ is nonnegative over $\mathscr{P}$ if and only if $M(\alpha^2)$ is nonnegative over the unit sphere $\tilde{\mathscr{D}}$ defined as follows:

$$\tilde{\mathscr{D}} = \{\alpha | \alpha_1^2 + \cdots + \alpha_k^2 = 1\} \tag{10.38}$$

- Assume that $N(\alpha)$ is a homogeneous matrix polynomial of degree γ. One can write:

$$N(\beta) = \|\beta\|^{\gamma} \times N\left(\frac{\beta}{\|\beta\|}\right), \quad \forall \beta \neq 0 \tag{10.39}$$

  Since $\frac{\beta}{\|\beta\|} \in \tilde{\mathscr{D}}$, one can conclude that $N(\alpha)$ is nonnegative over $\tilde{\mathscr{D}}$ if and only if it is nonnegative over the whole space $\mathbf{R}^k$.

Thus, the proof is completed by pursuing the argument given in the proof of Theorem 2, and using the inequalities (10.35) and (10.37) after replacing α with α^2 and removing the constraint $\alpha \in \mathscr{P}$ (note that the polynomial in the left side of the inequality (10.35) is homogeneous). ■

Optimization 2 *Maximize μ subject to the constraint that there exist a homogeneous matrix polynomial $P(\alpha)$ and SOS matrix polynomials $S_1(\alpha)$ and $S_2(\alpha)$ such that:*

$$P(\alpha^2) = S_1(\alpha), \tag{10.40a}$$

$$\begin{aligned}&\left[A(\alpha^2)\left(P(\alpha^2) + \mu(\alpha\alpha^T)^{\zeta_3} I_n\right) + \left(P(\alpha^2) + \mu(\alpha\alpha^T)^{\zeta_3} I_n\right) A^T(\alpha^2)\right] \\ &\times (\alpha\alpha^T)^{\max(0, 2\zeta_2 - \zeta_1 - \zeta_3)} + \tilde{B}(\alpha^2)\tilde{B}(\alpha^2)^T (\alpha\alpha^T)^{\max(0, \zeta_3 + \zeta_1 - 2\zeta_2)} = S_2(\alpha)\end{aligned} \tag{10.40b}$$

where ζ_3 denotes the degree of the polynomial $P(\alpha)$. Denote the solution of this optimization problem with $\tilde{\mu}^*$.

The following lemma is required in order to delve into the properties of the optimal parameter $\tilde{\mu}^*$ defined above.

Lemma 2 *Let $M(\alpha)$ be a homogeneous matrix polynomial with the property that $M(\alpha^2)$ is positive definite for every $\alpha \in \mathbf{R}^k \backslash \{0\}$. There exists a natural number c so that $(\alpha\alpha^T)^c M(\alpha^2)$ is SOS.*

Proof: The proof of this lemma relies heavily on the extension of Polya's theorem [18] to the matrix case, as carried out in [21]. More precisely, since $M(\alpha)$ is positive definite over the polytope $\mathscr{P}$, it follows from Theorem 3 in [21] that there exists a natural number c such that $(\alpha_1 + \alpha_2 + \cdots + \alpha_k)^c M(\alpha)$ has only positive-semidefinite matrix coefficients. This implies that the coefficients of $(\alpha\alpha^T)^c M(\alpha^2)$ are all positive-semidefinite, and in addition, its monomials are squared terms. As a result, $(\alpha\alpha^T)^c M(\alpha^2)$ is SOS. ■

Theorem 4 *The quantity* $\min_{\alpha\in\mathscr{P}}\{W_c(\alpha)\}$ *is equal to* $\tilde{\mu}^*$.

Proof: The proof will be performed in two steps. First, observe that if (10.40) holds for some matrices $P(\alpha)$, $S_1(\alpha)$ and $S_2(\alpha)$, then (10.32) is satisfied for $\tilde{P}(\alpha) = P(\alpha)$.

Conversely, assume that a matrix polynomial $\tilde{P}(\alpha)$ satisfies the inequalities given in (10.32), with the non-strict inequalities replaced by strict inequalities for every $\alpha \in \mathbf{R}^k \backslash \{0\}$ (such strict inequalities correspond to the case when the term *maximum* is substituted by *supremum*). Note that for a strict inequality in (10.32a), one would need to replace $\tilde{P}(\alpha)$ and μ with $\tilde{P}(\alpha) - \varepsilon(\alpha\alpha^T)^{\zeta_3}$ and $\mu + \varepsilon$, respectively, for a positive infinitesimal number ε. Now, one can apply Lemma 2 to these inequalities to conclude that there exists a natural number c such that if the expressions in the left sides of the inequalities (10.32a) and (10.32b) are multiplied by $(\alpha\alpha^T)^c$, then they become SOS matrix polynomials. It is enough to choose $P(\alpha)$ as $\tilde{P}(\alpha)(\alpha\alpha^T)^c$ for the inequalities given in (10.40) to hold (for some appropriate matrices $S_1(\alpha)$ and $S_2(\alpha)$). ■

Remark 2 *Optimization 2 is obtained from Theorem 3 by replacing the nonnegativity constraints with SOS conditions. Although one may argue that this can potentially introduce some conservatism due to the known gap between the set of SOS polynomials and the set of nonnegative polynomials, Theorem 4 shows that this is not the case. In other words, the proposed replacement does not make the resultant conditions conservative at all.*

Note that Optimization 2 can be handled using proper software as noted in Remark 1, provided some *a priori* upper bounds are set on the degrees of the relevant polynomials being sought. The question arises as to whether choosing higher degree polynomials would result in a tighter (less conservative) solution. This question is addressed in the next corollary.

Corollary 2 *The solution of Optimization 2 is a monotone nondecreasing function with respect to* ζ_3 *(the degree of the polynomial* $\tilde{P}(\alpha)$*).*

Proof: The proof is a direct consequence of the fact that if $\tilde{P}(\alpha)$ satisfies the constraints of Optimization 2 for some μ, then $\tilde{P}(\alpha)(\alpha_1 + \cdots + \alpha_k)$ also satisfies them for the same μ, but for some other suitable matrices $S_1(\alpha)$ and $S_2(\alpha)$. ■

10.3.2 Comparison with existing results

It follows immediately from the result of the paper [25] that the Gramian matrix can be approximated by a polynomial. Nonetheless, the relationship between the order of the corresponding polynomial and the approximation error is not investigated in [25]. In contrast, it is shown in Theorem 1 of the present chapter that the approximation error reduces exponentially with respect to the order of the corresponding polynomial. In particular, given a desired error bound, one can find the required degree of the approximating polynomial by computing μ_1 and μ_2 which depend on the smallest and largest eigenvalues of $A(\alpha)$ over the region $\mathscr{P}$. It is worth noting that the exponential reduction of the error is only for the approximation of the Gramian rational matrix with a polynomial matrix. Hence, for any fixed order of the polynomial approximation, Optimization 1 is to be solved in which the degrees of the polynomials $S_0(\alpha),\ldots,S_z(\alpha),\tilde{S}_0(\alpha),\ldots,\tilde{S}_z(\alpha)$ may be undesirably large. This issue arises in almost all optimization problems derived based on Polya's or Putinar's theorems.

As far as the complexity is concerned, it is easy to verify that Optimization 2 introduced in the present work is basically as complex as the optimization problem tackled in [3]. This partly results from the fact that there are an SOS homogeneous polynomial and two SOS constraints of a particular form in both approaches. For a detailed comparison between the complexities of the results given in [3], [2] and [1], the interested reader may refer to [1].

A recent paper [26] deals with the robust analysis and synthesis of linear uncertain systems. Although this paper presents useful results in the LMI framework, Optimization 2 is a much simpler problem compared to the ones proposed in [26]. It is noteworthy that all of the papers surveyed above deal with an inequality of the form:

$$A(\alpha)P(\alpha)+P(\alpha)A(\alpha)^T+\tilde{B}(\alpha)\tilde{B}(\alpha)^T \leq 0 \tag{10.41}$$

whereas this work studies the opposite inequality (see Theorem 1):

$$A(\alpha)P(\alpha)+P(\alpha)A(\alpha)^T+\tilde{B}(\alpha)\tilde{B}(\alpha)^T \geq 0 \tag{10.42}$$

This is a consequence of the fact that it is desired to *maximize* the *minimum* singular value. In other words, due to the special structure of the optimization considered here, the existing results cannot be directly applied to the underlying problem.

10.4 Numerical example

Example 1

A simple example is given here to profoundly illustrate the results of this work. Consider an LTI uncertain fourth-order system with the following state-space matrices:

$$A(\alpha) = \begin{bmatrix} -0.65 & -0.6 & -1 & 1.5 \\ -0.05 & -0.95 & -1.85 & -0.4 \\ -0.7 & 1.6 & -3.05 & 0.2 \\ -1.5 & -0.1 & -2.85 & -0.35 \end{bmatrix} \alpha_1 + \begin{bmatrix} -0.75 & 1.4 & 0 & 2.3 \\ -1.35 & -1.75 & -0.65 & -1.1 \\ 0.8 & 2.2 & -3.25 & -2.5 \\ -0.5 & 1.7 & 0.15 & 0.45 \end{bmatrix} \alpha_2,$$

$$B(\alpha) = \begin{bmatrix} -1.2 & -0.3 & -0.85 & 1.55 \\ 0.35 & -1.05 & -1.3 & -1.8 \\ 0.25 & 1.2 & -2.5 & -0.8 \\ -0.3 & 1.45 & -1 & -0.3 \end{bmatrix} \alpha_1 + \begin{bmatrix} 0.45 & 1.7 & 0.85 & 0.75 \\ -1.7 & -0.7 & 0.65 & 0.7 \\ 0.55 & 1 & -0.75 & -1.7 \\ -0.2 & 0.25 & 1.15 & 0.75 \end{bmatrix} \alpha_2 \tag{10.43}$$

where α_1 and α_2 are the uncertain parameters of the system, which belong to the polytope $\mathscr{P} = \{(\alpha_1, \alpha_2) | \alpha_1 + \alpha_2 = 1,\ \alpha_1, \alpha_2 \geq 0\}$. Regard (10.43) as an interconnected system with four inputs. It is desired to determine which inputs contribute weakly to the control of the system, and hence can be ignored for the sake of cost reduction. In other words, the objective is to find out which inputs play a vital role in controlling the system. To this end, for any given set $g \subset \{1,2,3,4\}$ let $\mathscr{S}^g(\alpha)$ represent the system $\mathscr{S}(\alpha)$ after ignoring those inputs whose indices belong to g. Denote the controllability Gramian of this system with $W_c^g(\alpha)$.

Given $\alpha \in \mathscr{P}$ and a final state x_0 of unit norm, consider the problem of finding an input $u(t)$ over the time interval $(-\infty, 0]$ with minimum L_2 norm such that it drives the state of the system $\mathscr{S}^g(\alpha)$ from $x(-\infty) = 0$ to $x(0) = x_0$. As discussed in Section 4.3 of [6], one possible solution is given by:

$$u_{opt}(t) = B^g(\alpha)^T e^{-A(\alpha)^T t} W_c^g(\alpha)^{-1} x_0, \quad t \leq 0 \tag{10.44}$$

where $B^g(\alpha)$ is the B-matrix of the system $\mathscr{S}^g(\alpha)$. The optimal input energy can be computed as:

$$\|u_{opot}\|^2 = x_0^T W_c^g(\alpha)^{-1} x_0 \tag{10.45}$$

Define $\nu(g)$ to be the maximum value of this optimal input energy over all final states x_0 of unit norm and all $\alpha \in \mathscr{P}$. The idea behind this definition is that $\nu(g)$ provides an upper bound for the input energy required to drive the state of the uncertain system $\mathscr{S}^g(\alpha)$, $\alpha \in \mathscr{P}$, from the origin at time $t = -\infty$ to any arbitrary point in the unit ball at time $t = 0$. Note that if $\nu(g)$ corresponds to a final state x_0^* and an uncertain parameter α^*, then x_0^* must be a unit eigenvector of $W_c^g(\alpha^*)$ associated with its smallest eigenvalue (singular value). This relationship can be expressed by:

$$\frac{1}{\nu(g)} = \min_{\alpha \in \mathscr{P}} \underline{\sigma}\{W_c^g(\alpha)\} \tag{10.46}$$

The objective is to evaluate $\nu(g)$ for different choices of g. To this end, four cases are considered as follows:

- *Case 1:* $g = \{2,3\}$.
- *Case 2:* $g = \{1\}$.
- *Case 3:* $g = \{4\}$.
- *Case 4:* $g = \{\}$.

To obtain $\nu(g)$ for any of the above cases, it suffices to solve Optimization 2 with the appropriate matrix $B^g(\alpha)$. For this purpose, the order of the polynomial $P(\alpha)$ being sought should be chosen *a priori*. This optimization is treated using YALMIP on a Dell laptop with a 1.6 GHz processor and 512 MB memory, and the results are given in Table 10.1. The last column of this table gives the points (α_1^*, α_2^*) for which the largest optimal input energy $\nu(g)$ is required. These points are computed by gridding the polytope properly, and performing an exhaustive search. Therefore, the entries of the last column are computed using a brute force technique, which will be exploited to verify the results obtained by solving Optimization 2. The second column of the table gives the solution of Optimization 2 at the second relaxation, i.e., when $P(\alpha)$ is assumed to be a homogeneous polynomial of order 2 (with the monomials α_1^2, α_2^2, $\alpha_1\alpha_2$). It can be verified that the solution obtained for any of the cases 1, 2 or 3 corresponds to the minimum singular value of the Gramian matrix evaluated at the optimal point given in the last column of the table. This proves that the relaxation arrives at the correct solution for cases 1, 2, and 3. For case 4, Optimization 2 is also solved at the fourth relaxation (by considering the monomials α_1^4, $\alpha_1^3\alpha_2$, $\alpha_1^2\alpha_2^2$, $\alpha_1^1\alpha_2^3$, α_2^4 for $P(\alpha)$). The corresponding solution is given in the third column, which is, in fact, the exact optimal value. The CPU time consumed for solving Optimization 2 at the second relaxation is given in the fourth column for each case; these values show that the problem is solved very fast. Using the results in columns 2 and 3 of the table as well as the equation (10.46), the quantity $\nu(g)$ is calculated and provided in column 5. It is to be noted that the proposed approach requires that the matrix $A(\alpha)$ be stable (Hurwitz) over the polytope. This can be verified using the method developed in [3]. It is worth mentioning that the ratio $\frac{\mu_1}{\mu_2}$ is equal to 0.075 in this example.

Table 10.1 Numerical results for Example 1

Case	Second relaxation	Fourth relaxation	CPU time for second relaxation	$\nu(g)$	(α_1^*, α_2^*)
1	0.0068	*	0.88 sec	147.06	(0.820, 0.180)
2	0.0744	*	0.82 sec	13.44	(0.240, 0.760)
3	0.1920	*	0.78 sec	5.21	(0.225, 0.775)
4	0.2516	0.2531	0.85 sec	3.97	(0.234, 0.766)

The values given in Table 10.1 (case 2) imply that the first input of the system is fairly important and ignoring it in the controller design would substantially increase the control energy required for shifting the state vector from certain points in the state-space. In contrast, the last input may be neglected, because its contribution is not significant as reflected by the small minimum singular value (case 3). However, if the second and third inputs are ignored concurrently (case 1), although the system remains robustly controllable by the remaining inputs, the required control energy would be huge.

For each of the above-mentioned cases, let the optimal input corresponding to the worst-case scenario (i.e. the final state x_0^* and the uncertain variable α^*) be applied to the system. Note that this input is given by (10.44), with $\alpha = \alpha^*$ (provided in Table 10.1) and $x_0 = x_0^*$ as defined earlier. The resulting input and state of the system are plotted in Figures 10.1, 10.2, 10.3, and 10.4. Notice that although these signals extend from $t = -\infty$ to $t = 0$, they are sketched only on the interval $t \in [-15, 0]$. These figures confirm the theoretical results obtained in this work. For instance, one can observe that in the case when both inputs $u_2(t)$ and $u_3(t)$ are blocked, the worst-case optimal input of the system has a large overshoot occurring at $t = 0$ (about 22 in magnitude).

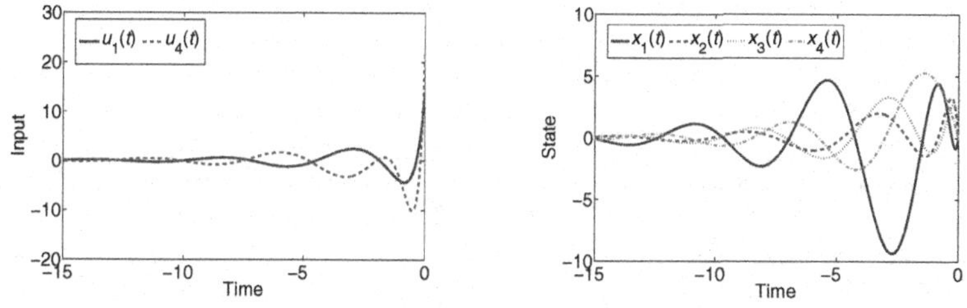

Fig. 10.1 The input and state of the system in case 1 (i.e., when inputs 2 and 3 are blocked).

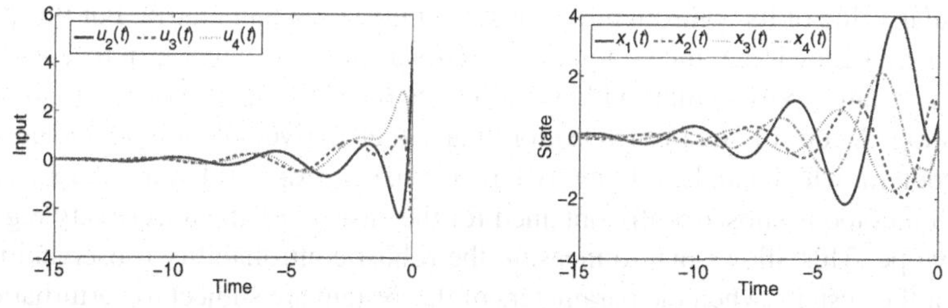

Fig. 10.2 The input and state of the system in case 2 (i.e., when input 1 is blocked).

10.5 Summary

Given a continuous-time linear time-invariant (LTI) system which is polynomially uncertain on a semi-algebraic region, this chapter obtains the minimum of the small-

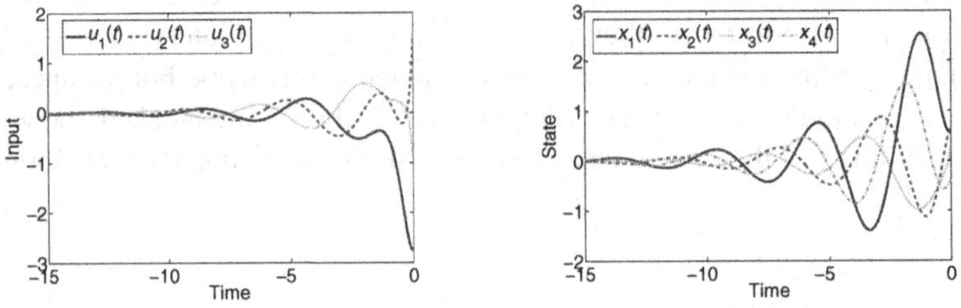

Fig. 10.3 The input and state of the system in case 3 (i.e., when input 4 is blocked).

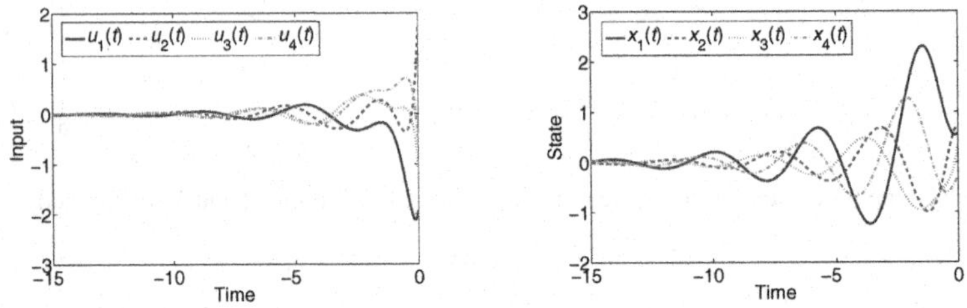

Fig. 10.4 The input and state of the system in (i.e., when none of the inputs is blocked).

est singular value of its controllability (observability) Gramian matrix. For this purpose, it is first shown that the Gramian is a rational function which can be approximated by a matrix polynomial (with any arbitrary precision) that satisfies an important relation. A sum-of-squares (SOS) formula is then derived for solving the underlying problem, which can be efficiently handled using proper software. An alternative SOS method is subsequently obtained for the case when the uncertainty region is a polytope. This allows one to measure the robust controllability (observability) degree of the system, when the parameters of the system are subject to perturbation. The method proposed here can be used to find a dominant subset of inputs and outputs for any given large-scale system, by determining the effectiveness of each input and output in the overall operation of the control system. Simulations demonstrate the effectiveness of the proposed results.

References

1. J. Lavaei and A. G. Aghdam, "Robust stability of LTI systems over semi-algebraic sets using sum-of-squares matrix polynomials," *IEEE Transactions on Automatic Control*, vol. 53, no. 1, pp. 417-423, 2008.
2. R. C. L. F. Oliveira and P. L. D. Peres, "LMI conditions for robust stability analysis based on polynomially parameter-dependent Lyapunov functions," *Systems & Control Letters*, vol. 55, no. 1, pp. 52-61, 2006.
3. G. Chesi, A. Garulli, A. Tesi, and A. Vicino, "Polynomially parameter-dependent Lyapunov functions for robust stability of polytopic systems: an LMI approach," *IEEE Transactions on Automatic Control*, vol. 50, no. 3, pp. 365-370, 2005.
4. S. Kau, Y. Liu, L. Hong, C. Lee, C. Fang, and L. Lee, "A new LMI condition for robust stability of discrete-time uncertain systems," *Systems & Control Letters*, vol. 54, no. 12, pp. 1195-1203, 2005.
5. M. C. de Oliveira and J. C. Geromel, "A class of robust stability conditions where linear parameter dependence of the Lyapunov function is a necessary condition for arbitrary parameter dependence," *Systems & Control Letters*, vol. 54, no. 11, pp. 1131-1134, 2005.
6. G. E. Dullerud and F. Paganini, A course in robust control theory: A convex approach, Texts in Applied Mathematics, Springer, 2005.
7. A. V. Savkin and I. R. Petersen, "Weak robust controllability and observability of uncertain linear systems," *IEEE Transactions on Automatic Control*, vol. 44, no. 5, pp. 1037-1041, 1999.
8. V. A. Ugrinovskii, "Robust controllability of linear stochastic uncertain systems," *Automatica*, vol. 41, no. 5, pp. 807-813, 2005.
9. S. S. Sastry and C. A. Desoer, "The robustness of controllability and observability of linear time-varying systems," *IEEE Transactions on Automatic Control*, vol. 27, pp. 933-939, 1982.
10. E. J. Davison and T. N. Chang, "Decentralized stabilization and pole assignment for general proper systems," *IEEE Transactions on Automatic Control*, vol. 35, no. 6, pp. 652-664, 1990.
11. D. D. Šiljak, Decentralized control of complex systems, Cambridge: Academic Press, 1991.
12. J. Lavaei and A. G. Aghdam, "A graph theoretic method to find decentralized fixed modes of LTI systems," *Automatica*, vol. 43, no. 12, pp. 2129-2133, 2007.
13. J. Lavaei and A. G. Aghdam, "Control of continuous-time LTI systems by means of structurally constrained controllers," *Automatica*, vol. 44, no. 1, pp. 141-148, 2008.
14. S. Sojoudi and A. G. Aghdam, "Characterizing all classes of LTI stabilizing structurally constrained controllers by means of combinatorics," in *Proceedings of 46th IEEE Conference on Decision and Control*, New Orleans, USA, 2007.
15. S. H. Wang and E. J. Davison, "On the stabilization of decentralized control systems," *IEEE Transactions on Automatic Control* vol. 18, no. 5, pp. 473-478, 1973.
16. G. Blekherman, "There are significantly more nonnegative polynomials than sums of squares," *Israel Journal of Mathematics*, vol. 153, no. 1, pp. 355-380, 2006.
17. P. A. Parrilo, "Structured semidefinite programs and semialgebraic geometry methods in robustness and optimization," *Ph.D. dissertation*, California Institute of Technology, 2000.
18. G. H. Hardy, J. E. Littlewood, and G. Polya, Inequalities. Cambridge University Press, Cambridge, UK, Second edition, 1952.
19. M. Putinar, "Positive polynomials on compact semi-algebraic sets," *Indiana University Mathematics Journal*, vol. 42, pp. 969-984, 1993.
20. J. Lavaei and A. G. Aghdam, "Optimal periodic feedback design for continuous-time LTI systems with constrained control structure," *International Journal of Control*, vol. 80, no. 2, pp. 220-230, 2007.
21. C. W. Scherer and C. W. J. Hol, "Matrix sum-of-squares relaxations for robust semi-definite programs," *Mathematical Programming*, vol. 107, no. 1-2, pp. 189-211, 2006.
22. J. Löfberg, "A toolbox for modeling and optimization in MATLAB," in *Proceedings of the CACSD Conference*, Taipei, Taiwan, 2004 (available online at http://control.ee.ethz.ch/$\sim$joloef/yalmip.php).

23. S. Prajna, A. Papachristodoulou, P. Seiler and P. A. Parrilo, "SOSTOOLS sum of squares optimization toolbox for MATLAB," *Users guide*, 2004 (available online at http://www.cds.caltech.edu/sostools).
24. C. J. Hillar and J. Nie, "An elementary and constructive solution to Hilbert's 17th Problem for matrices," in *Proceedings of the American Mathematical Society*, vol. 136, pp. 73-76, 2008.
25. P.-A. Bliman, R. C. L. F. Oliveira, V. F. Montagner, and P. L. D. Peres, "Existence of homogeneous polynomial solutions for parameter-dependent linear matrix inequal- ities with parameters in the simplex," in *Proceedings of 45th IEEE Conference on Decision and Control*, San Diego, USA, pp. 1486-1491, 2006.
26. R. C. L. F. Oliveira, M. C. de Oliveira, and P. L. D. Peres, "Convergent LMI relaxations for robust analysis of uncertain linear systems using lifted polynomial parameter-dependent Lyapunov functions," *Systems & Control Letters*, vol. 57, no. 8, pp. 680-689, 2008.

Chapter 11
Interconnection-Based Performance Analysis for a Class of Decentralized Controllers

11.1 Introduction

Many real-world systems can be described by large-scale interconnected models [1]. There is a great deal of interest in performance analysis and control synthesis of large-scale systems. A key practical consideration in designing a controller for this type of system is to rely on local information as much as possible. Decentralized control theory was introduced in the literature to address this consideration, and reduce the complexity of the control implementation for large-scale systems. Distinctive aspects of decentralized control systems have been well-documented in the last three decades [2, 3]. A decentralized controller consists of a number of isolated local controllers corresponding to the subsystems of the large-scale system. For the sake of simplicity of the control design problem, it is often desirable that the large-scale system possesses a hierarchical structure [4, 5]. Note that a hierarchical model refers to an interconnected system whose subsystems can be renumbered in such a way that the corresponding transfer function matrix becomes lower block-triangular. The control design problem for a hierarchical system can be broken down into a number of parallel design subproblems corresponding to different subsystems. The advantage of such design techniques is twofold: the control design procedure is far simpler for a number of low-order subsystems compared to that for one high-order system, and at the same time parallel computation is very fast.

Many important physical cooperative control applications with a leader-follower configuration such as formation flight, underwater vehicles, automated highway systems, satellite constellation, etc. have a hierarchical structure [6, 7, 8]. Furthermore, it is shown in [9] that under certain conditions, a continuous-time non-hierarchical system can have a hierarchical discrete-time equivalent model (this model represents the continuous-time system only at the sampling time instants). It is straightforward to show that a set of stabilizing local controllers obtained by neglecting all the interconnections between the subsystems constitute a stabilizing decentralized controller for the original hierarchical system. This is quite beneficial in the sense that it provides a simple control design method for hierarchical systems, as far as stability is concerned. In addition, a technique is given in [10] to design a near-optimal decentralized controller for hierarchical systems. This idea is further developed in [11] to decentralize any given centralized controller while its fundamental properties are preserved. Although decentralized control design for hierarchical systems has been extensively investigated in the past several years, there are only a few control design techniques for general large-scale systems, due to the complexity of the problem. Furthermore, there is no efficient performance evaluation method when the controller is designed for the system after some structural modifications.

On the other hand, there exist numerous non-hierarchical systems which are "close" to being hierarchical. Such systems have a few weak interconnections between their subsystems, whose elimination would result in an exact hierarchical structure. This type of systems will be referred to as *pseudo-hierarchical systems* throughout this chapter, and the hierarchical model obtained by eliminating minimum number of "weak" interconnections in a pseudo-hierarchical system will be called the corresponding *reference hierarchical model.* Given a pseudo-hierarchical large-scale system, a decentralized controller can be obtained for the corresponding reference hierarchical model by designing each local controller separately using available techniques. Even though this straightforward approach is appealing as far as the computational complexity is concerned, the decentralized controller obtained will not necessarily meet the design specifications for the original pseudo-hierarchical system. In fact, it may not even stabilize the original system, although it stabilizes the corresponding reference hierarchical model. Apart from the stability issue (which is not a concern if the interconnections which are removed to obtain the reference hierarchical model are sufficiently weak), the performance of the pseudo-hierarchical system under this controller can be quite poor. Thus, it is important to carry out a proper performance analysis for the system in order to make certain that this indirect design technique is suitable for the given pseudo-hierarchical large-scale system.

This chapter deals with the performance analysis for decentralized large-scale systems. It is assumed that a decentralized controller is provided for the reference hierarchical model of a pseudo-hierarchical system which meets certain control objectives. Moreover, it is supposed that the above-mentioned controller stabilizes the pseudo-hierarchical system, although closed-loop performance can be poor. A proper LQ cost function is defined to assess the discrepancy between the pseudo-hierarchical system and the corresponding reference hierarchical model under this decentralized controller. The smaller this performance index is, the closer the two closed-loop systems are to each other. Obtaining this cost function involves solving a discrete Lyapunov equation. However, due to the large-scale nature of the system, this equation might be difficult to handle, in general. Alternatively, it would be very useful to attain an upper bound on this cost function. A novel technique is proposed to address this objective, and it is subsequently shown that a LMI optimization problem with only three variables needs to be solved in order to compute this bound. This problem is also simplified and an explicit bound is proposed, without having to solve any optimization problem. One of the distinguishing features of this work is that it presents a simple technique for performance evaluation of pseudo-hierarchical decentralized systems. To elucidate that the obtained bounds are not too conservative in general, it is proved that if the difference between the pseudo-hierarchical system and the corresponding reference hierarchical model is sufficiently close to zero, so are these bounds. In particular, if the original model is exactly hierarchical, then both bounds will be equal to zero.

Another application of this work is in the control of overlapping systems, which has drawn much attention in recent years [12, 13, 14, 15]. Note that an overlapping system refers to a collection of subsystems, some of which share part of their states. For such a system, it is known that the most effective control structure is an overlapping one, in which any pair of local controllers can share data only if their corresponding subsystems share their states. The most prevalent method for designing an overlapping controller is to employ the inclusion principle to convert the problem to the conventional decentralized framework [16]. This is achieved by expanding the system into another system whose interconnections are all weak. Hence, a set of local controllers (which will altogether be referred to as a decentralized controller) can be designed for the individual subsystems of the expanded system by neglecting the interconnections. The overall decentralized controller will eventually be contracted to attain a decentralized overlapping controller for the original system. Since all the interconnections in the expanded system are weak, and moreover they are disregarded in the controller design, a proper performance analysis for this design technique would be very useful. To this end, the *expanded system* can be considered as a *pseudo-hierarchical model*, and the *expanded system with neglected interconnections* can be regarded as the corresponding *reference hierarchical model*. Thus, the performance analysis for the expanded system will fit into the formulation considered here, as a special case.

11.2 Preliminaries and problem formulation

Consider a large-scale interconnected system $\mathscr{S}$ consisting of ν subsystems, where its i-th subsystem S_i is represented by:

$$\begin{aligned} x_i[k+1] &= \sum_{j=1}^{\nu} A_{ij}x_j[k] + B_i u_i[k] \\ y_i[k] &= C_i x_i[k], \quad i \in \bar{\nu} := \{1,2,\cdots,\nu\} \end{aligned} \tag{11.1}$$

In the above equation, $x_i[k] \in \mathbb{R}^{n_i}$, $u_i[k] \in \mathbb{R}^{m_i}$ and $y_i[k] \in \mathbb{R}^{r_i}$ stand for the state, input and output of S_i, respectively. Sketch now a digraph $\mathscr{G}$ associated with the system $\mathscr{S}$ as follows:

- Assign ν vertices, one for each subsystem of $\mathscr{S}$.
- For any $i, j \in \bar{\nu}$, $i \neq j$, connect vertex i to vertex j with a directed edge if $A_{ij} \neq 0$.
- For any $i, j \in \bar{\nu}$, if there is an edge between vertex i and vertex j, attribute the weight $\|A_{ij}\|_F$ to that edge, where $\|\cdot\|_F$ denotes the Frobenius norm operator.

The graph $\mathscr{G}$ specifies the topology of information transfer between the subsystems. From this perspective, it plays an important role in the stability and stabilizability analysis of the system. If the graph $\mathscr{G}$ has no directed cycles, then the system $\mathscr{S}$ is said to be acyclic or hierarchical [10]. For any $i \in \bar{\nu}$, define the isolated subsystem $\bar{S}_i$ as:

$$\begin{aligned} \bar{x}_i[k+1] &= A_{ii}\bar{x}_i[k] + B_i\bar{u}_i[k] \\ \bar{y}_i[k] &= C_i\bar{x}_i[k] \end{aligned} \tag{11.2}$$

For the case when the graph $\mathscr{G}$ is acyclic, a stabilizing decentralized controller can be obtained by designing ν local controllers separately such that the i-th local controller stabilizes the isolated subsystem $\bar{S}_i$, for all $i \in \bar{\nu}$. This simple fact implies that when the graph $\mathscr{G}$ is acyclic, the decentralized controller design can be quite straightforward (as far as the stability is concerned). As noted earlier, several methods are proposed in the literature to design a LTI decentralized controller for a hierarchical system in order to achieve any pre-specified objectives [10, 5].

In the general case, when the graph $\mathscr{G}$ is not acyclic, one can remove certain edges of $\mathscr{G}$ to obtain an acyclic graph, and design the local controllers for the resultant system, as described before. However, the controller obtained may not perform satisfactorily when applied to the original system, if the interconnections neglected in the control design are not sufficiently weak. It is stated in [17] that there exist numerous systems for which all the edges of the graph $\mathscr{G}$ can be removed and designs local controllers accordingly. Nevertheless, in order to convert the graph $\mathscr{G}$ to an acyclic one, only a subset of the edges are required to be removed (such a subset is not necessarily unique).

Note that the ε-decomposition technique developed in the literature proposes a systematic method for eliminating certain interconnections of the system whose strength is below a specific level (see [18, 19, 20]). One can adopt above technique to identify the links whose elimination is not detrimental, and could lead to a hierarchical model.

Assume now that some of the edges are removed to obtain hierarchical model $\mathscr{S}_h$, and that a LTI decentralized controller $\mathscr{K}$ is designed for the resultant model using the existing methods. Once this controller is applied to the original system $\mathscr{S}$, the closed-loop system may perform poorly, and may even be unstable. Therefore, it is desired to evaluate the performance of the system $\mathscr{S}$ under the controller $\mathscr{K}$, with respect to its hierarchical counterpart (i.e. the hierarchical model under the same controller $\mathscr{K}$). To this end, it is assumed that the closed-loop system is stable, which is a requirement for performance degradation analysis in this work. It is worth mentioning that the closed-loop stability is guaranteed if the interconnections neglected in control design procedure are sufficiently weak.

In the sequel, the hierarchical model $\mathscr{S}_h$ under the designed LTI decentralized controller $\mathscr{K}$ is represented as:

$$x_h[k+1] = A_h x_h[k] \tag{11.3}$$

and the original system $\mathscr{S}$ under the same controller as:

$$x_c[k+1] = A_c x_c[k] \tag{11.4}$$

(it is important to note that both (11.3) and (11.4) are closed-loop equations). Furthermore, set $x_c[0] = x_h[0]$. With no loss of generality, the matrix A_h can be assumed to be lower-block triangular.

Remark 1 *In the case when a decentralized overlapping controller is to be designed for an overlapping system by mean of the inclusion principle, the expanded system (obtained from the original overlapping system) under the decentralized controller designed after neglecting all the interconnections is expressed by (11.4). The expanded system with nullified interconnections under the above-mentioned decentralized controller can then be described by (11.3), as a special case of a hierarchical model (this point is clarified in Example 1).*

In order to assess the closeness of the systems given in (11.3) and (11.4), one can measure the discrepancy between the states $x_h[k]$ and $x_c[k]$. This can be evaluated by the following performance index:

$$J_d = \sum_{k=0}^{\infty} (x_c[k] - x_h[k])^T (x_c[k] - x_h[k]) \tag{11.5}$$

Definition 1: Define the performance indices J_c and J_h as:

$$J_c = \sum_{k=0}^{\infty} x_c[k]^T x_c[k], \quad J_h = \sum_{k=0}^{\infty} x_h[k]^T x_h[k] \tag{11.6}$$

Definition 2: Given a positive real value μ, the controller $\mathscr{K}$ is said to be μ-suboptimal, if the inequality $\frac{J_d}{J_h} < \mu$ holds.

Some works such as [17] define the degree of suboptimality based on the ratio $\frac{J_c}{J_h}$, as opposed to $\frac{J_d}{J_h}$. However, it is manifest that the smallness of $\frac{J_c}{J_h}$ does not necessarily imply the closeness of $x_c[k]$ and $x_h[k]$. The objective here is to obtain a proper and easy-to-compute μ by which the controller $\mathscr{K}$ is suboptimal. The following practical restrictions are imposed.

Assumption 1: Given the system $\mathscr{S}$, a discrete Lyapunov equation with the order of the system (i.e. $n = \Sigma_{i=1}^{\nu} n_i$) cannot be solved efficiently due to the large-scale nature of the system, whereas a discrete Lyapunov equation with the order of any associated subsystem $\mathscr{S}_i$, $i \in \bar{\nu}$, can be handled more efficiently.

Assumption 2: Although solving a Lyapunov equation of order n is cumbersome, lower and upper bounds on the eigenvalues of a matrix of order n may be obtained efficiently.

It is important to note in Assumption 2 that in general solving a Lyapunov equation of order n is much more difficult than estimating the eigenvalues of a matrix of order n. In fact, the former problem involves n^2 variables while the latter one includes $n+1$ variables only (regardless of their linearity or bilinearity).

It is evident that J_h in (11.6) satisfies the relation:

$$J_h = x_h[0]^T P_h x_h[0] \tag{11.7}$$

where:

$$A_h^T P_h A_h - P_h + I = 0 \tag{11.8}$$

In order to obtain the main results of this chapter, one more assumption is required to be made.

Assumption 3: The closed-loop system given in (11.4) is stable with the Lyapunov matrix P_h.

It is to be noted that Assumption 3 is more restrictive than just the stability condition for the system (11.4), and is met when the removed edges have sufficiently small weights. Various sufficient conditions are provided in the literature to ensure the validity of this assumption.

11.3 Main results

In what follows, the performance deviation J_d will be formulated.

Lemma 1 *The performance index J_d is equal to:*

$$\begin{bmatrix} x_h[0]^T & x_h[0]^T \end{bmatrix} P_d \begin{bmatrix} x_h[0] \\ x_h[0] \end{bmatrix} \tag{11.9}$$

where:

$$\begin{bmatrix} A_h & 0 \\ 0 & A_c \end{bmatrix} P_d \begin{bmatrix} A_h & 0 \\ 0 & A_c \end{bmatrix}^T - P_d + \begin{bmatrix} I & -I \\ -I & I \end{bmatrix} = 0 \tag{11.10}$$

Proof: Augmenting the closed-loop systems (11.3) and (11.4) results in:

$$\begin{bmatrix} x_h[k+1] \\ x_c[k+1] \end{bmatrix} = \begin{bmatrix} A_h & 0 \\ 0 & A_c \end{bmatrix} \begin{bmatrix} x_h[k] \\ x_c[k] \end{bmatrix} \tag{11.11}$$

On the other hand, the performance index J_d can be rewritten as:

$$J_d = \sum_{k=0}^{\infty} \begin{bmatrix} x_h[k]^T & x_c[k]^T \end{bmatrix} \begin{bmatrix} I & -I \\ -I & I \end{bmatrix} \begin{bmatrix} x_h[k] \\ x_c[k] \end{bmatrix} \tag{11.12}$$

It is well-known that the above expression can be written as (11.9), where the matrix P_d satisfies the equation (11.10). This completes the proof. ■

Due to Assumption 1, the performance deviation J_d cannot be directly computed from Lemma 1 in order to compute the ratio $\frac{J_d}{J_h}$ precisely. Hence, the notion of μ-optimality is used here in order to obtain a reasonable upper bound on this ratio, which is carried out in the sequel.

Lemma 2 *Given a matrix H of proper dimension, assume the following inequality is satisfied:*

$$\begin{bmatrix} A_h & 0 \\ 0 & A_c \end{bmatrix}^T H \begin{bmatrix} A_h & 0 \\ 0 & A_c \end{bmatrix} - H + \begin{bmatrix} I & -I \\ -I & I \end{bmatrix} < 0 \tag{11.13}$$

Then, the inequality given below holds:

$$J_d < \begin{bmatrix} x_h[0]^T & x_h[0]^T \end{bmatrix} H \begin{bmatrix} x_h[0] \\ x_h[0] \end{bmatrix} \tag{11.14}$$

Proof: It can be concluded from the relations (11.10) and (11.13) that:

$$\begin{bmatrix} A_h & 0 \\ 0 & A_c \end{bmatrix}^T (H - P_d) \begin{bmatrix} A_h & 0 \\ 0 & A_c \end{bmatrix} - (H - P_d) < 0 \tag{11.15}$$

Since both of the matrices A_c and A_h are assumed to be Schur, it results from the above inequality that $P_d < H$. The proof follows immediately from this result and the equation (11.9). ■

Now, let the following optimization problem be introduced, which will be used later to show the μ-suboptimality of the controller.

Problem 1: Find the infimum of the objective function $k_1 + 2k_2 + k_3$ for the variables k_1, k_2 and k_3, subject to:

$$\begin{bmatrix} (1-k_1)I & k_2(A_h^T P_h A_c - P_h) - I \\ k_2(A_c^T P_h A_h - P_h) - I & k_3(A_c^T P_h A_c - P_h) + I \end{bmatrix} < 0 \tag{11.16}$$

Remark 2 *Problem 1 is a LMI optimization which can be efficiently handled using proper software tools such as YALMIP or SOSTOOLS [21, 22].*

Theorem 1 *The controller $\mathscr{K}$ is μ-suboptimal, where μ denotes the infimum obtained by solving Problem 1.*

Proof: Consider any real scalars k_1, k_2 and k_3 satisfying the inequality (11.16) given in Problem 1. Using the equation given in (11.8), this inequality can be rewritten as:

$$\begin{bmatrix} A_h & 0 \\ 0 & A_c \end{bmatrix}^T H \begin{bmatrix} A_h & 0 \\ 0 & A_c \end{bmatrix} - H + \begin{bmatrix} I & -I \\ -I & I \end{bmatrix} < 0 \tag{11.17}$$

if the matrix H is chosen as:

$$H = \begin{bmatrix} k_1 P_h & k_2 P_h \\ k_2 P_h & k_3 P_h \end{bmatrix} \tag{11.18}$$

Therefore, it can be inferred from Lemma 2 that:

$$\begin{aligned} J_d &< \begin{bmatrix} x_h[0]^T & x_c[0]^T \end{bmatrix} H \begin{bmatrix} x_h[0] \\ x_c[0] \end{bmatrix} = (k_1 + 2k_2 + k_3) x_h[0]^T P_h x_h[0] \\ &= (k_1 + 2k_2 + k_3) J_h \end{aligned} \tag{11.19}$$

(note that $x_h[0] = x_c[0]$). Thus:

$$\frac{J_d}{J_h} < k_1 + 2k_2 + k_3 \tag{11.20}$$

The above inequality implies that the minimum of the objective function $k_1+2k_2+k_3$ is to be minimized, and this completes the proof. ∎

Theorem 1 states that the solution of Problem 1 provides an upper bound on the ratio $\frac{J_d}{J_h}$. It is interesting to note that the inequality constraint of this optimization problem is always feasible. To prove this, it suffices to choose $k_1=2$, $k_2=0$ and let k_3 be a very large number. Note from Assumption 3 that the matrix $A_c^T P_h A_c - P_h$ is negative definite; hence, the inequality (11.16) holds.

Due to Assumption 1 and the large-scale nature of the system $\mathscr{S}$, Problem 1 may not be easily solvable in practice. This is mainly because of the matrix constraint (11.16) which becomes sophisticated for large-scale systems. Thus, it is desirable to convert the matrix inequality (11.16) into a scalar form. This objective will be addressed in the sequel.

Problem 2: Find the infimum of the objective function $k_1+2k_2+k_3$ for the variables k_1,k_2 and k_3 subject to the scalar inequalities $k_1>1$ and:

$$(k_1-1)(-1+k_3m_1)-1-k_2^2m_2-k_2m_3>0 \tag{11.21}$$

where:

$$\begin{aligned} m_1 &= \underline{\lambda}(-R_2), \quad m_2=\bar{\lambda}\left(R_1R_1^T\right), \\ m_3 &= \bar{\lambda}\left(-R_1-R_1^T\right), \\ R_1 &= A_c^T P_h A_h - P_h, \quad R_2 = A_c^T P_h A_c - P_h \end{aligned} \tag{11.22}$$

(the notations $\bar{\lambda}(\cdot)$ and $\underline{\lambda}(\cdot)$ represent the maximum and minimum magnitudes of the eigenvalues of a matrix, respectively).

Theorem 2 *Denote with μ the infimum obtained by solving Problem 2. Then, the controller $\mathscr{K}$ is μ-suboptimal.*

Proof: The inequality constraint of Problem 1 can be rearranged as:

$$\begin{bmatrix} (k_3R_2+I) & (k_2R_1-I) \\ (k_2R_1^T-I) & (1-k_1)I \end{bmatrix} < 0 \tag{11.23}$$

Applying the Schur complement formula to the above inequality results in:

$$(k_1-1)(k_3R_2+I)+(k_2R_1-I)\left(k_2R_1^T-I\right)<0 \tag{11.24}$$

It is easy to verify that the matrix inequality (11.24) is guaranteed to hold, provided the scalar inequality given below is satisfied:

$$\underline{\lambda}\big((k_1-1)(-k_3R_2-I)\big) > \bar{\lambda}\left((I-k_2R_1)\left(I-k_2R_1^T\right)\right) \tag{11.25}$$

Choose some scalars k_1,k_2,k_3 satisfying the inequality (11.21). It can be deduced from the above discussion and the result of Theorem 1 that in order to prove Theorem 2 it suffices to substantiate the validity of the inequality (11.25). To this end, one can use the following equation:

$$\underline{\lambda}\big((k_1-1)(-k_3R_2-I)\big) = (k_1-1)(-1+k_3m_1) \tag{11.26}$$

Moreover, it results from Lemma 2.1 in [23] that:

$$\begin{aligned}\bar{\lambda}\left(\left(I-k_2R_1\right)\left(I-k_2R_1^T\right)\right) &= 1+\bar{\lambda}\left(k_2\left(-R_1-R_1^T\right)+k_2^2R_1R_1^T\right)\\ &\leq 1+k_2\bar{\lambda}\left(-R_1-R_1^T\right)+k_2^2\bar{\lambda}\left(R_1R_1^T\right)\leq 1+k_2m_3+k_2^2m_2\end{aligned} \tag{11.27}$$

The relations (11.21), (11.26) and (11.27) altogether lead to the inequality (11.25). ■

Remark 3 *Similar to the previous case, it can be shown that the constraints of Problem 2 are always feasible (by considering $k_1 = 2$, $k_2 = 0$ and a sufficiently large number for k_3). It is to be noted that m_1 is positive by definition.*

Remark 4 *Since the statement of Problem 2 is obtained by reducing the matrix constraint in Problem 1 to some scalar constraints, the upper bound proposed for μ in Theorem 2 is more conservative than the one given in Theorem 1.*

To solve any of the two problems introduced in this chapter, the Lyapunov matrix P_h needs to be obtained first. As a consequence of Assumption 1, this matrix cannot be computed efficiently using the conventional methods. However, since the matrix A_h (which is required for obtaining P_h) is assumed to be lower block-triangular, P_h can be found by solving a number of Lyapunov and Sylvester equations of subsystems' orders (as opposed to the system's order), successively. This fact is elaborated in detail in the appendix.

Theorem 3 *Denote the optimal values of the variables k_1, k_2, k_3 in Problem 2 with k_1^*, k_2^*, k_3^*. The triple (k_1^*, k_2^*, k_3^*) satisfies either the set of equations:*

$$k_1^* = 1 \tag{11.28a}$$

$$I - k_2^* R_1 = 0 \tag{11.28b}$$

$$k_3^* = \frac{1}{m_1} \tag{11.28c}$$

or the following ones:

$$(4m_2^2 - 4m_1m_2)(k_2^*)^2 + (4m_2m_3 - 4m_1m_3)k_2^* + (m_3^2 - 4m_1) = 0 \tag{11.29a}$$

$$k_3^* = \frac{-2m_2k_2^* + m_3}{2m_1} + \frac{1}{m_1} \tag{11.29b}$$

$$k_1^* = \frac{m_2(k_2^*)^2 + m_3k_2^* + 1}{k_3^*m_1 - 1} + 1 \tag{11.29c}$$

Proof: Since there is a strict inequality constraint in Problem 2 (namely $k_1 > 1$), the Karush-Kuhn-Tucker (KKT) method cannot be employed here (in fact, if this constraint is replaced by $k_1 \geq 1$ to enable the exploitation of the KKT method, a wrong result may be obtained in which $k_3^* = -\infty$). Therefore, let all different possibilities be investigated below:

- Case 1: *The equation* $k_1^* - 1 = 0$ *holds.* In this case, the underlying optimization problem reduces to finding the infimum of $1 + 2k_2 + k_3$ under the constraint $-1 - k_2^2 m_2 - k_2 m_3 > 0$. Since the resultant objective function has no local minimum as noted above, the optimal solution occurs at some point on the remaining boundary, i.e. $-1 - (k_2^*)^2 m_2 - k_2^* m_3 = 0$. On the other hand, it follows from (11.27) that:

$$0 \leq \bar{\lambda}\big((I - k_2^* R_1)\,(I - k_2^* R_1^T)\big) \leq 1 + k_2^* m_3 + (k_2^*)^2 m_2 \tag{11.30}$$

The above relation along with the equation $-1 - (k_2^*)^2 m_2 - k_2^* m_3 = 0$ signifies that the matrix $I - k_2^* R_1$ is equal to zero. Taking this result into account, it can be concluded from the constraint of Problem 2 and the equation $k_1^* = 1^+$ that $k_3^* m_1 - 1$ is nonnegative. This implies that in order for the objective function to be minimized, k_3^* should be chosen as $\frac{1}{m_1}$. The relations obtained above satisfy the set of equations given in (11.28).
- Case 2: *The equation* $(k_1^* - 1)(-1 + k_3^* m_1) - 1 - (k_2^*)^2 m_2 - k_2^* m_3 = 0$ *holds.* It can be inferred in this case that:

$$k_1^* = \frac{m_2 (k_2^*)^2 + m_3 k_2^* + 1}{k_3^* m_1 - 1} + 1 \tag{11.31}$$

If $k_1^* - 1$ is equal to zero, this case turns out to be the same as case 1. Hence, with no loss of generality, assume that $k_1^* - 1$ is strictly positive. This yields that solving Problem 2 is equivalent to finding the lowest minimum point of the function:

$$\frac{m_2 k_2^2 + m_3 k_2 + 1}{k_3 m_1 - 1} + 1 + 2k_2 + k_3 \tag{11.32}$$

for which k_1^* obtained in (11.31) is greater than or equal to 1. Taking the gradient of the above function and equating it to zero will lead to the equations:

$$\frac{m_2 (k_2^*)^2 + m_3 k_2^* + 1}{(k_3^* m_1 - 1)^2} \times (-m_1) + 1 = 0 \tag{11.33a}$$

$$2 + \frac{2 m_2 k_2^* + m_3}{m_1 k_3^* - 1} = 0 \tag{11.33b}$$

One can combine these two equations to arrive at the relation (11.29a). The proof follows from the fact that the equations (11.33b) and (11.31) are identical to (11.29b) and (11.29c), respectively.
- Case 3: *The strict inequalities* $(k_1^* - 1)(-1 + k_3^* m_1) - 1 - (k_2^*)^2 m_2 - k_2^* m_3 > 0$ *and* $K_1^* > 1$ *both hold.* As shown in case 1, the relation $k_3^* m_1 - 1 \geq 0$ must hold. This implies that there exists a sufficiently small positive number η such that the triple $(k_1^* - \eta, k_2^*, k_3^*)$ satisfies the constraints of Problem 2, while the value of the objective function at this triple is smaller than its minimum value. This contradiction rules out this case. ∎

Theorem 3 presents a solution to Problem 2, which in turn (according to Theorem 2) provides a value for μ (i.e., the suboptimality degree of the controller $\mathscr{K}$). Regarding the set of equations (11.29) in this theorem, one should note that the quadratic equation (11.29a) needs to be solved first. The solution should then be substituted into the equations (11.29b) and (11.29c) to find all other parameters.

The question arises, how conservative are the values of μ obtained in Theorems 1 and 2? To answer this question, an elegant result on the tightness of this bound will be presented next.

Theorem 4 *In the case where A_h and A_c are identical, Theorems 1 and 2 both arrive at the exact solution $\mu = 0$.*

Proof: If $A_h = A_c$, then it can be easily verified that $R_1 = R_2 = -I$, and consequently $m_1 = m_2 = 1$ and $m_3 = 2$. Now, Theorem 2 states (after some simplifications) that μ is equal to the infimum of $k_1 + 2k_2 + k_3$ under the inequality constraints $k_1 > 1$ and:

$$k_1k_3 - k_1 - k_3 - 2k_2 - k_2^2 > 0 \tag{11.34}$$

The latter inequality is equivalent to:

$$(k_1 - 1)(k_3 - 1) > (k_2 + 1)^2 \tag{11.35}$$

Hence, μ is equal to 0, and is attained when $k_1 = k_3 \to 1^+$ and $k_2 = -1$. ■

Remark 5 *It can be inferred from Theorem 4 and the continuity, that if A_c is sufficiently close to A_h, then the upper bounds proposed in this chapter will be arbitrarily close to zero. As can be noticed from the proof of this theorem, the result is not trivial at all. In other words, it is not straightforward to conclude from Theorems 1 and 2 that if $A_c = A_h$, then the corresponding upper bounds will be equal to the exact value, implying $\mu = 0$.*

Remark 6 *The results obtained in this work can be analogously developed to tackle the following problem:*
Assume that the system $\mathscr{S}$ (which is not necessarily hierarchical) is subject to perturbation. Design a LTI decentralized controller for the nominal model of the system. Now, the matrices A_h and A_c correspond to the closed-loop nominal and the closed-loop perturbed A-matrices in the state space representation, respectively. In this case, the ratio $\frac{J_d}{J_c}$ describes the closeness of the nominal closed-loop system and its perturbed counterpart. As in the previous problem, this ratio would be very useful in evaluating the performance of a physical system under a controller designed for the nominal model of the system.

11.4 Numerical example

Example 1: Consider an overlapping system $\mathscr{S}$ given by:

$$x[k+1] = Ax[k] + Bu[k] \tag{11.36}$$

where:

$$A = \left[\begin{array}{c|c|c} 1 & 0 & 0 \\ 0.1 & 0.3 & -0.1 \\ 0 & 1 & 0 \end{array}\right], \quad B = \left[\begin{array}{c|c} 0.5 & 0 \\ 0 & 0 \\ 0 & 0.5 \end{array}\right] \tag{11.37}$$

with the overlapping part A_{22}. The objective is to design an overlapping controller for this system with the control law $u[k] = \mathscr{K} x[k]$, which minimizes the performance index $J = \sum_0^\infty x[k]^T x[k] + u[k]^T u[k]$. Notice that the matrix $\mathscr{K}$ to be designed must have the following structure:

$$\mathscr{K} = \begin{bmatrix} * & * & 0 \\ 0 & * & * \end{bmatrix} \tag{11.38}$$

A prevailing approach to achieve this objective is to exploit the expansion-contraction method [16]. Using a transformation matrix V given by:

$$V = \begin{bmatrix} 1 & 0 & 0 \\ 0 & 1 & 0 \\ 0 & 1 & 0 \\ 0 & 0 & 1 \end{bmatrix} \tag{11.39}$$

an expanded system $\mathscr{S}_e$ can be obtained as follows [24, 25]:

$$x_e[k+1] = A_e x_e[k] + B_e u_e[k] \tag{11.40}$$

where:

$$A_e = \begin{bmatrix} 1 & 0 & 0 & 0 \\ 0.1 & 0.3 & 0 & -0.1 \\ 0.1 & 0 & 0.3 & -0.1 \\ 0 & 0 & 1 & 0 \end{bmatrix}, \quad B_e = \begin{bmatrix} 0.5 & 0 \\ 0 & 0 \\ 0 & 0 \\ 0 & 0.5 \end{bmatrix} \tag{11.41}$$

A controller $u_e[k] = \mathscr{K}_e x_e[k]$ is now to be designed for the system $\mathscr{S}_e$ to minimize the performance index $J_e = \sum_0^\infty x_e[k]^T x_e[k] + u_e[k]^T u_e[k]$, where, according to the overlapping technique, the matrix $\mathscr{K}_e$ must have the following form:

$$\mathscr{K}_e = \begin{bmatrix} * & * & 0 & 0 \\ 0 & 0 & * & * \end{bmatrix} \tag{11.42}$$

To simplify the control design problem, one can use the overlapping technique to replace the matrix A_e with the following:

$$\bar{A}_e = \begin{bmatrix} 1 & 0 & 0 & 0 \\ 0.1 & 0.3 & 0 & 0 \\ 0 & 0 & 0.3 & -0.1 \\ 0 & 0 & 1 & 0 \end{bmatrix} \tag{11.43}$$

Consequently, the optimal decentralized controller for the expanded system $\mathscr{S}_e$ (using the above matrix) is given by:

$$\mathscr{K}_e = \begin{bmatrix} -0.7867 & -0.0037 & 0 & 0 \\ 0 & 0 & -0.3997 & -0.0022 \end{bmatrix} \tag{11.44}$$

It is worth mentioning that since $\bar{A}_e$ is a block-diagonal matrix, it represents two decoupled subsystems, and hence designing $\mathscr{K}_e$ is reduced to finding two separate controllers, which is easier in general. The corresponding near-optimal controller for the original system $\mathscr{S}$ is:

$$\mathscr{K} = \begin{bmatrix} -0.7867 & -0.0037 & 0 \\ 0 & -0.3997 & -0.0022 \end{bmatrix} \tag{11.45}$$

Since the matrix $\mathscr{K}_e$ is designed for the system $\mathscr{S}_e$ with the parameters $(\bar{A}_e, B_e)$ instead of (A_e, B_e), it is desired to verify how this would affect the performance of the closed-loop system. For this purpose, consider the matrices A_h and A_{nh} as $\bar{A}_e + B_e\mathscr{K}_e$ and $A_e + B_e\mathscr{K}_e$, respectively. Using Theorem 1, one can easily obtain $k_1 = 1.1909, k_2 = -1.1955$ and $k_3 = 1.2240$, which leads to a suboptimal controller with the performance index 0.024. The smallness of this number confirms that neglecting the interconnections in the proposed design procedure does not result in a noticeable performance degradation. It is to be noted that the exact value of $\frac{J_d}{J_h}$ in this example is equal to 0.0116. The result of this example is in accordance with the discussion in Remark 1.

Example 2: Consider an interconnected system $\mathscr{S}$ with nine SISO subsystems of order 1, and assume that the interconnections from subsystem i to subsystem j, for $i, j \in \{1, 2, ..., 9\}$, $i < j$, are in general "weaker" than the ones in the opposite direction. Hence, to design a decentralized controller for the system with nine local controllers, one can eliminate these weak interconnections and design a decentralized controller for the obtained hierarchical model using any existing method. For simplicity, assume that a static decentralized controller has been designed for the hierarchical model. To carry out performance analysis for the pseudo-hierarchical system under the designed controller, two different choices will be considered for the closed-loop matrix A_{nh} in the sequel.

Consider first a matrix A_{nh} of the following form:

$$\frac{1}{4.35}\begin{bmatrix} 1 & 0.5 & 2 & 0.1 & 0.5 & 0.6 & 0.3 & 0.3 & 0.1 \\ 0 & 1 & 1.5 & 0.5 & 1 & 0 & 1 & 0.2 & 0.25 \\ 1 & 0.3 & 1 & 0 & 0.2 & 1 & 0.2 & 0.5 & 0.31 \\ 0 & 0 & 0.3 & 1 & 3 & 1 & 0.05 & 0.1 & 0.01 \\ 0.3 & 0 & 0 & 1 & 1 & 2 & 0 & 0 & 0.2 \\ 0 & 0 & 0 & 0 & 1 & 1 & 0.8 & 0 & 1 \\ 0 & 0 & 0.04 & 0.5 & 0.6 & 0 & 0.5 & 1 & 1 \\ 0.01 & 0 & 0 & 0.1 & 0.1 & 0 & 0.5 & 1 & 2 \\ 0.4 & 0.9 & 0.04 & 0.03 & 0 & 0.3 & 0.05 & 1 & 0 \end{bmatrix}$$

It can be observed that the lower-diagonal entries of this matrix have smaller magnitudes compared to the upper-diagonal ones in general (which translate to "weak" interconnections in the digraph of the system). The hierarchical matrix (A_h) obtained by neglecting the lower-diagonal entries of the above matrix is given by:

$$\frac{1}{4.35}\begin{bmatrix} 1 & 0.5 & 2 & 0.1 & 0.5 & 0.6 & 0.3 & 0.3 & 0.1 \\ 0 & 1 & 1.5 & 0.5 & 1 & 0 & 1 & 0.2 & 0.25 \\ 0 & 0 & 1 & 0 & 0.2 & 1 & 0.2 & 0.5 & 0.31 \\ 0 & 0 & 0 & 1 & 3 & 1 & 0.05 & 0.1 & 0.01 \\ 0 & 0 & 0 & 0 & 1 & 2 & 0 & 0 & 0.2 \\ 0 & 0 & 0 & 0 & 0 & 1 & 0.8 & 0 & 1 \\ 0 & 0 & 0 & 0 & 0 & 0 & 0.5 & 1 & 1 \\ 0 & 0 & 0 & 0 & 0 & 0 & 0 & 1 & 2 \\ 0 & 0 & 0 & 0 & 0 & 0 & 0 & 0 & 0 \end{bmatrix} \tag{11.46}$$

In can be verified that J_d and J_h for this example are equal to 67.9336 and 38.6145, respectively, and hence $\frac{J_d}{J_h} = 1.7593$. On the other hand, an upper bound on $\frac{J_d}{J_h}$ can be obtained from Theorem 1 by solving Problem 1, which results in:

$$k_1 = 2.131, \quad k_2 = -1.7529, \quad k_3 = 5.7894$$

From the relation $\mu = \min(k_1 + 2k_2 + k_3)$, the upper bound μ on the ratio $\frac{J_d}{J_h}$ is equal to 4.4145. Note that although the resultant upper bound is approximately 2.5 times greater than the exact value, it is attained through a quite simple procedure, which is very desirable for large-scale systems. This relatively large difference between $\frac{J_d}{J_h}$ and the corresponding upper bound μ is due to the fact that the neglected interconnections are not "weak" enough to be ignored. For instance, there are some large lower-triangular entries (such as $\frac{0.9}{4.35}$), which are comparable to and even greater than some of the upper-triangular entries.

Since Problem 1 involves matrix variables, it may not be easily handled for large-scale systems. Thus, let Theorems 2 and 3 be utilized here to find an upper bound on J_d/J_h. In this case, k_1, k_2 and k_3 are equal to 1.5206, -0.9144 and 5.2813, respectively, which lead to the upper bound $\mu = 4.973$. Although the matrix constraint in Problem 1 may seem to be oversimplified in obtaining Problem 2, the results of Example 2 show that this is not necessarily the case. In other words, the bound $\mu = 4.973$ obtained by using Theorem 3 is relatively close to the bound $\mu = 4.4145$ resulted from Theorem 1.

Example 3: Consider the matrix A_{nh} as follows:

$$\frac{1}{4.35}\begin{bmatrix} 1 & 0.5 & 2 & 0.1 & 0.5 & 0.6 & 0.3 & 0.3 & 0.1 \\ 0 & 1 & 1.5 & 0.5 & 1 & 0 & 1 & 0.2 & 0.25 \\ 0 & 0 & 1 & 0 & 0.2 & 1 & 0.2 & 0.5 & 0.31 \\ 0 & 0 & 0 & 1 & 3 & 1 & 0.05 & 0.1 & 0.01 \\ 0 & 0 & 0 & 0 & 1 & 2 & 0 & 0 & 0.2 \\ 0.05 & 0 & 0 & 0 & 0 & 1 & 0.8 & 0 & 1 \\ 0 & 0 & 0 & 0 & 0 & 0 & 0.5 & 1 & 1 \\ 0 & 0 & 0 & 0 & 0 & 0 & 0 & 1 & 2 \\ 0.001 & 0 & 0 & 0 & 0 & 0 & 0 & 0 & 0 \end{bmatrix}$$

Here, the "weak" interconnections of the previous system have been further weakened in order for the pseudo-hierarchical system to become closer to its reference hierarchical model. Notice that the reference hierarchical model of this system is the same as the one in the previous example, i.e. (11.46). It can be verified that J_d and J_h in this case are equal to 0.0117 and 38.6145, respectively. The substantial drop in the magnitude of J_d compared to the previous case, manifestly confirms that the closeness of the pseudo-hierarchical and the corresponding reference hierarchical models would have a significant impact on the effectiveness of the proposed performance evaluation. The upper bound μ obtained by using Theorems 1 and 2 are equal to 0.0041 and 0.0671, respectively.

11.5 Summary

This chapter deals with the performance analysis of large-scale systems with pseudo-hierarchical structures; i.e., those systems whose transfer function matrix is "close" to being hierarchical, due to some weak elements (interconnections) in it. It is assumed that a stabilizing decentralized controller is available for the system. This controller is basically designed for a reference hierarchical model; i.e., a hierarchical model which is obtained by eliminating some weak interconnections in the original system. The controller obtained by using this indirect design technique may not perform well when applied to the original pseudo-hierarchical system. A proper LQ cost function is defined to measure the discrepancy between the performance of the original system and its hierarchical counterpart, under the above controller. Since computing the exact value of this cost function involves a large-scale Lyapunov equation, in practice it is preferable to obtain an upper bound on it. Hence, a simple LMI optimization problem with only three variables is proposed to attain this upper bound. To further simplify the procedure, the matrix optimization problem is reduced to a scalar one for which an explicit solution is obtained. In addition, it is shown that the closer the pseudo-hierarchical system to the reference hierarchical model is, the smaller these bounds are. In the special case when the two models are identical, these bounds are both equal to zero. This demonstrates that the bounds obtained by using the simplified optimization problems are not too conservative. The usefulness of the proposed techniques is illustrated by three numerical examples.

11.6 Appendix

It is desired to show that since the matrix P_h is the solution of a Lyapunov equation involving a hierarchical matrix (A_h), it can alternatively be found by solving a number of Lyapunov and Sylvester equations of subsystems' orders (as opposed to the system's order), successively. To streamline the argument, assume that there are only two subsystems (i.e. $\nu = 2$). In this case, the equation $A_h^T P_h A_h - P_h + I = 0$ can be equivalently decomposed as:

$$A_{22}^T P_3 A_{22} - P_3 + I = 0 \tag{11.47a}$$
$$A_{11}^T P_2 A_{22} + A_{21}^T P_3 A_{22} - P_2 = 0 \tag{11.47b}$$
$$A_{11}^T P_1 A_{11} + A_{11}^T P_2 A_{21} + A_{21}^T P_2 A_{11} + A_{21}^T P_3 A_{21} - P_1 + I = 0 \tag{11.47c}$$

where:

$$P_h = \begin{bmatrix} P_1 & P_2 \\ P_2 & P_3 \end{bmatrix} \tag{11.48}$$

Since the hierarchical closed-loop system is stable, the matrices A_{11} and A_{22} are both Schur. Thus, the Lyapunov equation (11.47a) which is of subsystem's order, can be solved to find the matrix P_3. Substitution of P_3 in the equation (11.47b) will yield a Sylvester equation, which has a unique solution P_2 (because the eigenvalues of A_{11} and A_{22} are all inside the unit circle). Finally, the Lyapunov equation (11.47c) can be solved for the matrix variable P_1 after substituting P_2 and P_3 obtained above into this equation. This illustrates that due to the special structure of A_h, Assumption 1 is not essential for hierarchical systems. An analogous method can be adopted in the general case (i.e. $\nu > 2$), in order to obtain the Lyapunov matrix corresponding to the large-scale system through a series of Sylvester and Lyapunov equations of smaller sizes.

One may think that the above-mentioned *back substitution* method is potentially problematic, because solving a hierarchy of equations could lead to the propagation of the round-off error and eventually a numerical instability. However, the Gaussian elimination method for solving a linear upper-triangular system (which is a certain type of back substitution technique), is known to be numerically stable [26]. It can be similarly shown that the aforementioned method is numerically efficient too. To clarify this, let the (i, j) block entry of P_h be denoted by P_{ij}, and assume that solving the Sylvester equation associated with the variable P_{ij}, $i, j \in \bar{\nu}$, $i \leq j$, yields the solution $\tilde{P}_{ij}$, which is subject to round-off error. It can be verified that:

$$A_h^T \tilde{P} A_h - \tilde{P} = -I + A_d^T \Delta A_d - \Delta \tag{11.49}$$

where:

- $\tilde{P}$ is a matrix whose (i, j) block entry is $\tilde{P}_{ij}$, for all $i, j \in \bar{\nu}$.
- A_d is a block diagonal matrix with the (i,i) block entry A_{ii}, for all $i \in \bar{\nu}$.
- Δ is a matrix with the (i, j) block entry Δ_{ij}, for all $i, j \in \bar{\nu}$, where Δ_{ij} is a matrix of the same dimension as P_{ij}, representing the round-off error produced in the course of solving the Sylvester equation associated with P_{ij}.

The equation (11.49) somehow indicates that there does not exist any error propagation phenomenon, and as long as Δ is small, $\tilde{P}$ would be close to P_h. More precisely, one can write:

$$A_h^T (\tilde{P} - P_h) A_h - (\tilde{P} - P_h) = A_d^T \Delta A_d - \Delta \tag{11.50}$$

Since all entries of Δ are small (say, of order $O(\varepsilon)$) and A_d is block diagonal, one would expect to obtain a solution $\tilde{P} - P_h$ of small norm (see [27] and references therein for relating the norm of $\tilde{P} - P_h$ to the norm of Δ). This confirms that the proposed method is not prone to the propagation error, i.e. $\tilde{P}$ is normally close to P_h.

The other issue about the proposed back substitution method is that it significantly reduces the computational burden, provided the orders of the subsystems are sufficiently smaller than the order of the system. For instance, assume that all subsystems have roughly the same dimension, say $\frac{n}{\nu}$ $(n := n_1 + \cdots + n_\nu)$. Then, the computational complexity of the Lyapunov solution P_h using the existing efficient methods is $O(n^3)$ [28]. In contrast, by taking the particular structure of A_h into account, the present method requires that $\frac{\nu(\nu+1)}{2}$ Sylvester equations be solved, whose overall complexity order is $\nu^2 \times O((\frac{n}{\nu})^3)$. This implies that the proposed algorithm reduces the computational complexity by a factor of $\frac{1}{\nu}$.

References

1. A. C. Antoulas and D.C. Sorensen, "Approximation of large-scale dynamical systems: An overview," *International Journal of Applied Mathematics and Computer Science*, vol. 11, no. 5, pp. 1093-1121, 2001.
2. D. D. Šiljak, Decentralized Control of Complex Systems, Boston: Academic Press, 1991.
3. M. Jamshidi, Large-Scale Systems: Modeling, Control, and Fuzzy Logic, Prentice-Hall, NJ, 1997.
4. S. S. Stankovic and D. D. Šiljak, "Sequential LQG optimization of hierarchically structured systems," *Automatica*, vol. 25, no. 4, pp. 545-559, 1989.
5. J. Lavaei, A. Momeni and A. G. Aghdam, "A model predictive decentralized control scheme with reduced communication requirement for spacecraft formation," *IEEE Transactions on Control Systems Technology*, vol. 16, no. 2, pp. 268-278, 2008.
6. H. G. Tanner, G. J. Pappas and V. Kumar, "Leader to formation stability," *IEEE Transactions on Robotics and Automation*, vol. 20, no. 3, pp. 443-455, 2004.
7. J. A. Fax and R. M. Murray, "Information flow and cooperative control of vehicle formations," *IEEE Transactions on Automatic Control*, vol. 49, no. 9, pp. 1465-1476, 2004.
8. D. J. Stilwell and B. E. Bishop, "Platoons of underwater vehicles," *IEEE Control Systems Magazine*, vol. 20, no. 6, pp. 45-52, 2000.
9. A. G. Aghdam, E. J. Davison and R. Becerril, "Structural modification of systems using discretization and generalized sampled-data hold functions," *Automatica*, vol. 42, no. 11, pp. 1935-1941, 2006.
10. J. Lavaei, A. Momeni and A. G. Aghdam, "High-performance decentralized control for formation flying with leader-follower structure," in *Proceedings of 45th IEEE Conference on Decision and Control*, San Diego, CA, 2006.
11. J. Lavaei and A. G. Aghdam, "High-performance decentralized control design for general interconnected systems with applications in cooperative control," *International Journal of Control*, vol. 80, no. 6, pp. 935-951. 2007.
12. A. Iftar, "Overlapping decentralized dynamic optimal control," *International Journal of Control*, vol. 58, no. 1, pp. 187-209, 1993.
13. S. S. Stankovic, M. J. Stanojevic, and D. D. Šiljak, "Decentralized overlapping control of a platoon of vehicles," *IEEE Transactions on Control Systems Technology*, vol. 8, no. 5, pp. 816-832, 2000.
14. D. D. Šiljak and A. I. Zecevic, "Control of large-scale systems: Beyond decentralized feedback," *Annual Reviews in Control*, vol. 29, no. 2, pp. 169-179, 2005.

15. A. I. Zecevic and D. D. Šiljak, "A new approach to control design with overlapping information structure constraints," *Automatica*, vol. 41, no. 2, pp. 265-272, 2005.
16. D. Chu and D. D. Šiljak, "A canonical form for the inclusion principle of dynamic systems," *SIAM Journal on Control and Optimization*, vol. 44, no. 3, pp. 969-990, 2005.
17. R. Krtolica and D. D. Šiljak, "Suboptimality of decentralized stochastic control and estimation," *IEEE Transactions on Automatic Control*, vol. 25, no. 1, pp. 76-83, 1980.
18. M. E. Sezer and D. D. Šiljak "Nested ε-decompositions and clustering of complex systems," *Automatica*, vol. 22, no. 3, pp. 321-331, 1986.
19. M. E. Sezer and D. D. Šiljak "Nested epsilon decompositions of linear systems: Weakly coupled and overlapping blocks," *SIAM Journal of Matrix Analysis and Applications*, vol. 12, pp. 521-533, 1991.
20. A. I. Zecevic and D. D. Šiljak "A block-parallel Newton method via overlapping epsilon decompositions," *SIAM Journal of Matrix Analysis and Applications*, vol. 15, pp. 824-844, 1994.
21. J. Löfberg, "A toolbox for modeling and optimization in MATLAB," in *Proceedings of the CACSD Conference*, Taipei, Taiwan, 2004 (available online at http://control.ee.ethz.ch/~joloef/yalmip.php).
22. S. Prajna, A. Papachristodoulou, P. Seiler and P. A. Parrilo, "SOSTOOLS sum of squares optimization toolbox for MATLAB," *Users guide*, 2004 (available online at http://www.cds.caltech.edu/sostools).
23. C. H. Lee, "Solution bounds of the continuous and discrete Lyapunov matrix equations," *Journal of Optimization Theory and Applications*, vol. 120, no. 3, pp. 559-578, 2004.
24. L. Bakule, J. Rodellar and J. M. Rossell, "Inclusion principle for uncertain discrete-time systems with guaranteed cost," in *Proceedings of 43rd IEEE Conference on Decision and Control*, Atlantis, Paradise Island, Bahamas, 2004.
25. L. Bakule, J. Rodellar, J. M. Rossell and P. Rubió "Preservation of controllability-observability in expanded systems," *IEEE Transactions on Automatic Control*, vol. 46, no. 7, pp. 1155-1162, 2001.
26. L. N. Trefethen and D. Bau, "Numerical Linear Algebra," *SIAM*, 1997.
27. M. K. Tippett and D. Marchesin,"Upper bounds for the solution of the discrete algebraic Lyapunov equation," *Automatica*, vol. 35, no. 8, pp. 1485-1489, 1999.
28. P. Benner,"Large-scale matrix equations of special type," *Numerical Linear Algebra with Applications*, vol. 15, no. 9, pp. 747-754, 2008.

Index

affine variety, 143
algebraic variety, 143
approximate decentralized fixed modes, 82
automated highway systems, 171

bipartite graph, 15

centralized controller, 88
communication networks, 61
consensus, 109
control interaction set, 76
controllability Gramian, 152
cooperative control, 171

decentralized controller, 10
decentralized fixed mode, 10
decentralized implementation, 112
decentralized overlapping control, 37
decentralized overlapping fixed modes, 47
decentralized servomechanism controller, 88
derandomization technique, 26
deterministic algorithms, 24
digraph, 11

expanded system, 55

flight formation, 109

generalized sampled-data hold function, 50
geographically distributed subsystems, 139

hierarchical structure, 171

implementation cost, 82
information flow matrix, 37
information flow structure, 38
inherent cost, 82
interconnected system, 23
Internet congestion control, 109

Kalman observable form, 119
Kosaraju's algorithm, 29
Kronecker product, 48

large space structures, 61
large-scale systems, 172
leader-follower formation, 103
local controller, 25
LQR decentralized control, 110
Lyapunov equation, 172

matrix inequality, 93
maximal subgraphs, 66
minimal sets, 67

near-optimal decentralized controller, 88

observability Gramian, 152
observer-based controller, 118
optimal centralized controller, 88
optimal control interaction set, 81
optimal decentralized controller, 122
optimal overlapping structure, 62

performance index, 98
perturbed model, 100
polynomial uncertainties, 149
polynomially uncertain systems, 151
power systems, 61
pseudo-hierarchical system, 172

quotient fixed mode, 11
quotient overlapping fixed mode, 54
quotient system, 11

randomized algorithms, 24
robust control, 149
robust controllability, 160
robust controllability degree, 155
robust performance, 152

sampled-data controller, 51
separation principle, 121
servomechanism controller, 92
stabilizability, 61
stabilizing controller, 38
state estimation error, 120
strongly connected, 11
structural graph, 11
structural initial value observable, 118
structurally constrained controller, 40
structurally fixed modes, 146
structurally robust fixed mode, 143
structured uncertainty, 151
Sylvester matrix, 144

Tarjan's algorithm, 29
time complexity, 24
transmission zero, 48

unobservable subspace, 119
unstructured uncertainty, 151

vehicle formation, 88

weak interconnections, 172